T0323769

THE HEART OF THE WILD

The Heart of the Wild

ESSAYS ON NATURE, CONSERVATION,
AND THE HUMAN FUTURE

EDITED BY

BEN A. MINTEER AND
JONATHAN B. LOSOS

WITH ESSAYS BY

Bill Adams, Joel Berger, Susan Clayton, Eileen Crist, Martha
L. Crump, Thomas Lowe Fleischner, Harry W. Greene,
Hal Herzog, Jonathan B. Losos, Emma Marris, Ben A. Minteer,
Kathleen Dean Moore, Gary Paul Nabhan, Peter H. Raven,
Christopher J. Schell, Richard Shine, and Kyle Whyte

PRINCETON UNIVERSITY PRESS

PRINCETON & OXFORD

Published by Princeton University Press
41 William Street, Princeton, New Jersey 08540
99 Banbury Road, Oxford OX2 6JX

press.princeton.edu

Library of Congress Cataloging-in-Publication Data

Names: Minteer, Ben A., 1969– editor. | Losos, Jonathan B., editor.
Title: The heart of the wild : essays on nature, conservation, and the
    human future / edited by Ben A. Minteer and Jonathan B. Losos.
Description: Princeton : Princeton University Press, [2024] | Includes
    bibliographical references.
Identifiers: LCCN 2023042840 (print) | LCCN 2023042841 (ebook) |
    ISBN 9780691228624 (hardback) | ISBN 9780691228617 (ebook)
Subjects: LCSH: Nature. | Wildlife conservation. | Human ecology. |
    BISAC: NATURE / Essays | SCIENCE / Life Sciences / Biology
Classification: LCC QH81 .H395 2024 (print) | LCC QH81 (ebook) |
    DDC 508—dc23/eng/20231201
LC record available at https://lccn.loc.gov/2023042840
LC ebook record available at https://lccn.loc.gov/2023042841

British Library Cataloging-in-Publication Data is available

Editorial: Alison Kalett and Hallie Schaeffer
Production Editorial: Ali Parrington
Jacket Design: Karl Spurzem
Production: Danielle Amatucci
Publicity: Matthew Taylor and Kate Farquhar-Thomson
Copyeditor: Dana Henricks

Jacket image: Kevin Kopf / Shutterstock

CONTENTS

ABOUT THE CONTRIBUTORS

BILL (W. M.) ADAMS is Claudio Segré Professor of Conservation and Development, Geneva Graduate Institute, and Emeritus Moran Professor of Conservation and Development at the University of Cambridge. His research explores conservation from the perspectives of political ecology and environmental history, and he is currently interested in the ways in which new technologies reshape understanding of nature and conservation practice. Recent books include *Green Development: Environment and Sustainability in a Developing World* (4th ed., Routledge, London, 2020), and *Strange Natures: Conservation in the Era of Synthetic Biology* (with Kent Redford, Yale University Press, 2021).

JOEL BERGER is the Barbara Cox-Anthony Chair in Wildlife Conservation at Colorado State University and a Senior Scientist for the Wildlife Conservation Society. He's studied and published on endangered species in Namibia (black rhinos, mountain zebras, bats), Chile and Argentina (huemul), and rare or endangered ones in central Asia including saiga, takin, and wild yak. Berger has explored the remote and the wild at the planet's icy edges from the Arctic and sub-Antarctic and at high elevation in the Himalayas, and has written or coedited many books, including *Horn of Darkness: Rhinos on the Edge*; *The Better to Eat You With: Fear in the Animal World*; and *Extreme Conservation—Life at the Edges of the World*. He is a Fellow of the American Association for Advancement of Sciences, a multiple Guggenheim awardee, and the recipient of lifetime achievement awards from the Society of Conservation Biology and the American Society of Mammalogists.

SUSAN CLAYTON is Whitmore-Williams Professor of Psychology at the College of Wooster in Ohio. Her research, which has been conducted at several zoos around the world, focuses on the human relationship with nature, how it is socially constructed, and how it can be utilized to promote environmental conservation. She has also written extensively about the implications of climate change for human well-being. Her most recent book is *Psychology and Climate Change: Human Perceptions, Impacts, and Responses* (2018).

EILEEN CRIST taught in the Department of Science, Technology, and Society at Virginia Tech for twenty-two years, retiring in 2020. Her work focuses on the extinction crisis and loss of wild places, pathways to halt these trends, and inquiries surrounding humanity's relationship with the planet. She is coeditor of a number of books, including *Gaia in Turmoil: Climate Change, Biodepletion, and Earth Ethics in an Age of Crisis*, and *Keeping the Wild: Against the Domestication of Earth*. She has written numerous academic papers as well as popular writings and is Associate Editor of the online journal *The Ecological Citizen*. Her most recent book, *Abundant Earth: Toward an Ecological Civilization*, was published by the University of Chicago Press in 2019.

MARTHA L. CRUMP is adjunct professor in the departments of biology at Utah State University and Northern Arizona University. She is a behavioral ecologist and conservation biologist and has worked extensively with amphibians in Central and South America. She received the Distinguished Herpetologist Award from the Herpetologists' League (1997), the Henry S. Fitch Award for Excellence in Herpetology from the American Society of Ichthyologists and Herpetologists (2020), and is a past president of the Society for the Study of Amphibians and Reptiles. Her most recent book is *Women in Field Biology: A Journey into Nature*, coauthored with Michael J. Lannoo.

THOMAS LOWE FLEISCHNER was the founding director of the Natural History Institute, in Prescott, Arizona, and currently serves as its Senior Advisor and Director Emeritus. He is Faculty Emeritus at Prescott College, where he taught interdisciplinary environmental studies for twenty-nine years, Past Chair of the Natural History Section of the

Ecological Society of America, a Fellow of the Linnean Society of London, and author or editor of numerous articles and four books, the most recent being *Nature, Love, Medicine: Essays on Wildness and Wellness.*

HARRY W. GREENE was for four decades a professor at the University of California, Berkeley, and Cornell University, each of which honored him with campus-wide teaching awards. Now an adjunct professor at the University of Texas at Austin, he is a fellow of the American Academy of Arts and Sciences, and authored *Snakes: The Evolution of Mystery in Nature* and *Tracks and Shadows: Field Biology as Art.* These days Harry hangs out with Pancho, Lefty, and his other longhorns while writing a book about wildness.

HAL HERZOG has been studying the morally and psychologically complex relationships between people and animals for over thirty years. The author of over one hundred research publications, his articles have also appeared in media outlets such as the *New York Times,* the *Washington Post, Time Magazine,* and *Wired Magazine.* His book, *Some We Love, Some We Hate, Some We Eat: Why It's So Hard to Think Straight About Animals,* has been translated into ten languages. An award-winning teacher and researcher, he is Emeritus Professor of Psychology at Western Carolina University.

JONATHAN B. LOSOS is an evolutionary biologist known for his research on how lizards rapidly evolve to adapt to changing environments. He is the William H. Danforth Distinguished University Professor at Washington University and Director of the Living Earth Collaborative, a biodiversity partnership between the university, the Saint Louis Zoo, and the Missouri Botanical Garden. Losos has written more than 240 papers and three books, and is an author of a leading college biology textbook. He has been elected a member of the National Academy of Sciences and a fellow of the American Academy of Arts and Sciences and is the recipient of a Guggenheim Fellowship and many other awards.

EMMA MARRIS is an environmental writer based in Oregon. She is the author of *Wild Souls: Freedom and Flourishing in the Non-Human World* and *Rambunctious Garden: Saving Nature in a Post-Wild World.*

Emma also writes for *National Geographic,* the *New York Times,* *Nature,* and *Outside,* among other publications.

BEN A. MINTEER is professor of environmental ethics and conservation in the School of Life Sciences at Arizona State University. He has authored or edited more than a dozen books, including *Wild Visions: Wilderness as Image and Idea; The Fall of the Wild: Extinction, De-Extinction, and the Ethics of Conservation;* and *The Landscape of Reform: Civic Pragmatism and Environmental Thought in America.*

KATHLEEN DEAN MOORE is the author or coeditor of a dozen books about our moral relation with nature, among them *Great Tide Rising, Wild Comfort,* and *Moral Ground.* Her most recent book is *Earth's Wild Music: Celebrating and Defending the Songs of the Natural World.* Distinguished Professor of Philosophy Emerita at Oregon State University, she now writes and speaks widely about the moral urgency of climate action.

GARY PAUL NABHAN is a student of plant-animal-culture interactions in deserts, an ethnobiologist, contemplative ecologist, conservation biologist, and biocultural geographer trained at the University of Arizona and Prescott College. A MacArthur Fellow, he is the author or editor of more than thirty books translated into six languages, a number of which have won awards. In addition to his research, teaching, and community service on sustainable food systems, Nabhan was mentored by Harry Greene while proposing, writing and completing *Singing the Turtles to Sea,* the first book length ethnoherpetology, published by University of California Press. His most recent books include *Agave Spirits* (2023) and *The Nature of Desert Nature* (2020).

PETER H. RAVEN is a leading botanist and advocate of conservation and biodiversity, President Emeritus of the Missouri Botanical Garden, and George Engelmann Professor of Botany Emeritus at Washington University in St. Louis. For more than fifty years, he has been an outspoken advocate of the need for conservation throughout the world and has received numerous prizes and awards in recognition of these activities. Peter has authored or coauthored numerous books and publications, both popular and scientific, including *Biology of Plants*, the internationally best-selling textbook in botany, and *Environment*, a leading textbook on the environment. He has held both Guggenheim and MacArthur Fellowships.

CHRISTOPHER J. SCHELL is an urban ecologist whose research combines evolutionary theory with ecological applications to disentangle the processes accentuating human-carnivore conflict. His interdisciplinary work requires integrating principles from the natural sciences with urban studies to address how systemic racism and oppression affect urban ecosystems, while simultaneously highlighting the need to promote environmental justice, civil rights, and equity as the bedrock of biological conservation and our fight against the climate crisis. He is an Assistant Professor in the Department of Environmental Science, Policy, and Management at the University of California, Berkeley.

RICHARD SHINE is a Professor in Biology at Macquarie University in Sydney, Australia. His research aims to develop new approaches to conservation of ectothermic organisms, based on understanding their fundamental ecology. Author of the books *Cane Toad Wars, So Many Snakes So Little Time,* and *Snakes Without Borders,* Shine has published more than 1,000 scientific papers and has won many awards for his research, including the E. O. Wilson Naturalist Award and Robert Whittaker Distinguished Ecologist Award.

KYLE WHYTE is Professor of Environment and Sustainability and George Willis Pack Professor at the University of Michigan. Kyle's research addresses moral and political issues concerning climate policy and Indigenous peoples, the ethics of cooperative relationships between Indigenous peoples and science organizations, and problems of Indigenous justice in public and academic discussions of food sovereignty, environmental justice, and the Anthropocene. He is an enrolled member of the Citizen Potawatomi Nation. Kyle has partnered with numerous Tribes, First Nations, and inter-Indigenous organizations in the Great Lakes region and beyond on climate change planning, education, and policy.

Introduction

WILD HEARTS AND MINDS

Ben A. Minteer and Jonathan B. Losos

THE ARTIST Thomas Cole knew well what it meant to "civilize" the landscape. An émigré to the United States from Lancashire, he'd witnessed firsthand the force of industrial expansion as it transformed England's rural districts in the early nineteenth century. It was an experience he didn't wish to see repeated in his new home, a nation so closely identified with its seemingly endless wilderness. Arriving with his family in Philadelphia in 1818 at the age of seventeen, Cole soon moved to Pittsburgh, and then on to New York City. His time in New York was a turning point in his development as an artist, as Cole took his first trip up the Hudson River in the late summer of 1825 and saw the Catskill Mountains for the first time. Over the next several years, he explored the region's wildlands, all the while sketching and recording his experiences.

Cole's paintings of the Catskills captured the rough-hewn beauty of this rugged corner of the Northeast, a romantic depiction of the wilderness as a vast, powerful, yet ultimately beautiful place that offered not just a therapeutic retreat from industrial and commercial life, but the opportunity to commune with the transcendent. It was an aesthetic and cultural resource, Cole believed, that was increasingly imperiled in Jacksonian

America, a time of relentless economic and territorial expansion that left behind a devastating trail of ecological and social wreckage.

Cole was an artist, but he had a naturalist's eye and an empiricist's methodological instincts. He was one of the first painters in America to work directly from the field, and his interest in geology and natural history, as well as his emphasis on the critical importance of close observation, shaped the work that adorned his canvases. And those canvases quickly became some of the most prized works of American art ever produced, with Cole ascending to the role of the nation's premier landscape painter, father of the Hudson River School, and a powerful influence on a generation of prominent artists, from Asher Durand to Cole's student, Frederic Church.

Cole's excursions in the Catskills and New England's White Mountains also confronted him with the environmental consequences of early nineteenth-century economic development. He was increasingly distressed by the transformation of the countryside in the Northeast, especially the impacts of extensive timber harvesting in the 1830s. In 1836, Cole produced a remarkable series of paintings that dramatized his fear that Americans were courting disaster by "over-civilizing" the landscape and destroying the wilderness, what he and other nature romantics saw as the source of moral virtue and nationalistic pride.

The Course of Empire was a five-part sequence of canvases depicting the development of a single allegorical wild landscape, from a primordial state of nature, to a transitory pastoral scene, to the emergence of a gleaming imperial city. The mountain wilderness that dominates the image in the early paintings (fig. 1) gradually recedes and becomes partly obscured by the rise of civilization, a visual progression that conveys the passage of time and the march of "progress."

In the fourth painting in the series, Cole's visual narrative takes a dark turn. A swirling storm envelops the scene, which shows the formerly resplendent city being destroyed by an invading force, a violent and fiery sacking clearly meant to evoke the fall of Rome (fig. 2). The final painting in the sequence, *Desolation*, depicts the abandoned city in ruins, completing the cyclical historical view of the rise and fall of civilization. The mountain in the background reemerges more completely as the wilderness returns, reclaiming the broken city.

FIGURE 1. Thomas Cole, *The Course of Empire: The Savage State*, 1836. Oil on canvas, The New-York Historical Society. Source: Wikimedia Commons.

FIGURE 2. Thomas Cole, *The Course of Empire: Destruction*, 1836. Oil on canvas, The New-York Historical Society. Source: Wikimedia Commons.

Although *The Course of Empire* features a generic European (presumably Mediterranean) setting, it's clear that the American audience was very much on Cole's mind when he created the series. As he wrote that same year (1836):

> the most distinctive, and perhaps the most impressive, characteristic of American scenery is its wildness. It is the most distinctive, because in civilized Europe the primitive features of scenery have long since been destroyed or modified—the extensive forests that once overshadowed a great part of it have been felled—rugged mountains have been smoothed, and impetuous rivers turned from their courses to accommodate the tastes and necessities of a dense population—the once tangled wood is now a grassy lawn; the turbulent brook a navigable stream—crags that could not be removed have been crowned with towers, and the rudest valleys tamed by the plow.

Cole's paintings were thus not only masterfully realized artistic works, they were a warning to Americans that they, too, could suffer Europe's environmental fate. As Roderick Nash writes in his classic text, *Wilderness and the American Mind*, Cole feared that if Americans lost the vital connection to their wild roots, they would lose the natural source of virtue and independence upon which the Republic rested. The ethical, spiritual, and political consequences of falling too far from wild nature would, in other words, prove catastrophic. Cole's anxiety, and his aesthetic and moral argument for preserving American nature (and especially the wilderness), would course through contemporaneous and subsequent generations of American naturalists and conservationists in the late nineteenth and early twentieth centuries, from Ralph Waldo Emerson and Henry David Thoreau to John Muir, Teddy Roosevelt, and Aldo Leopold.

And yet the environmentalism expressed in works like *The Course of Empire* rested upon some assumptions that today seem problematic. Chief among them is the mythic sense of the wilderness as a sublime, "pristine," unpeopled place, a land out-of-time offering a cultural and political tabula rasa to a young nation claiming an environmentally and culturally rendered exceptionalism. It's a view that's been dismantled as we've accepted

FIGURE 3. Thomas Cole, *The Course of Empire: Desolation*, 1836. Oil on canvas, The New-York Historical Society. Source: Wikimedia Commons.

a more complex ecological and historical narrative, one defined instead by the flux of nature and the ubiquity of the human presence.

Cole's wilderness idealism and romanticism, that is, can appear naively antiquarian and even ethnocentric today, a melodramatic and preachy paean to a lost and imaginary world for a privileged audience. For those not inclined to ecological alarmism and, especially, for critics of the environmental movement, he becomes one of the first of a long line of Cassandras warning that doom awaits any society in the grip of its own hubris and afflicted by an insatiable hunger for its dwindling natural resources. And yet for others, *The Course of Empire* remains an unsettling and even prophetic work, a cultural and historical reminder that we haven't escaped Cole's sequence—or heeded its ecological and moral lessons (fig. 3).

———

"One of the penalties of an ecological education," observed the eminent conservationist and writer Aldo Leopold, "is that one lives alone in a world of wounds." Nowadays we might quibble with the "alone" part—the ranks of the ecologically educated have swelled significantly since Leopold wrote those words in the early 1940s—but the larger point still rings true. The ecological discords and losses that have inexorably followed the expanding human footprint on the planet weigh heavily on the hearts and minds of conservationists and nature lovers today. From mass extinctions and the loss of wildlands to the spread of ocean microplastics and global climate change, Leopold's "wounds" have never been more cutting, nor more worrying.

But as Leopold understood, the all-important desire to *act* in response to this awareness of ecological ills isn't only a matter of ecological education, of learning how we're disrupting biogeochemical cycles, rending ecological communities, pushing species off the evolutionary cliff, and destabilizing the climate. It's also a question of getting people to *connect* with the species, places, and nature experiences most vulnerable to our expanding activities and development pathways.

This connection, in turn, can lead to a deeper aesthetic, moral, and emotional bond with the natural world that augments the scientific understanding. Key to this process, Leopold believed (as have the generations of ecologists and conservationists traveling in his wake), is our ability to nourish and sustain a commitment to *naturalism*: a curiosity about biological diversity and natural history and an abiding concern for the wild as we push deeper into a human-shaped and increasingly technological and urban future. Articulating and defending this ethos, and understanding both its demands and delights, is the primary motivation of this book.

The Heart of the Wild is a meditation on the urgency of learning about, experiencing, and caring for nature and the wild in a time of expanding human impact. Our authors are a top-flight group of scientists, environmental studies scholars, and nature writers whose work and thought have advanced our understanding of the beauty, diversity, and vulnerability of nature (wild and otherwise) and our responsibility to learn from and sustain it. In this book, we asked them to assay the trends and forces—cultural, technological, and conceptual—that are reframing our understanding of nature and our relationship to it, including those that seem to be pulling us farther away from wild species and places each passing day.

Specifically, we invited our authors to respond to one or both of the following questions, which together serve as the organizing force binding this collection: What do we need to observe, experience, and value in nature and the wild as it changes under human influence in order to square our role within it, now and in the future? And how can we keep a love of nature and wild things alive in an increasingly human-defined age? Although each contributor brings a unique perspective, background, and voice to these questions, three main themes emerge from the essays that follow.

One is the rapidly changing landscape and shifting assumptions of biological conservation, a concern that drives the discussion in the book's first section ("Conservation's Shifting Ground"). The common message emerging from these essays is that we need to relinquish older notions of privileging native species in unaltered, unchanging, and "pristine" places, that is, those elements of nature orthodoxy that depict human activity as inherently polluting an otherwise spotless ecological order. Indeed, as Emma Marris writes, we need to understand instead how the human presence can be "woven into the web without destroying it." Rick Shine draws a similar conclusion, arguing that we must adapt to changing times and accept our ecological disturbances, our wounded wildernesses, for the beauty and knowledge they still offer us. Even Australia's notorious cane toad has a role to play in such reshuffled ecosystems. Dingoes, too: as one of us (Jonathan) writes in his essay, this "dog gone wild" compels us to both rethink our ecological assumptions and rearrange the targets of conservation for the Anthropocene. Hal Herzog's reflections on the wildness of domestic cats and the uncomfortable ethical questions raised by our commitment to these beloved predators remind us how our assumptions about the moral and evolutionary gulf between wild and domestic species quickly collapse under closer scrutiny.

In the final contribution to this section, Peter Raven broadens the temporal and spatial scope of the conversation with his sweeping account of our global biotic history. It is a story that drives home the ecological and social stakes of the current moment—not to mention the precariousness of our wild future.

The second set of essays ("Wilderness, Wildness, Wild: Legacies and Liabilities") explores the limitations and value of the idea of wilderness, and related conceptions of "wildness" and "wild," in environmental ethics and practice. Kyle Whyte opens the section by questioning the value today of appeals to wild concepts in the effort to understand places and species, especially given the painful history of wilderness preservation for Indigenous peoples. Eileen Crist counters with a strong defense of the wilderness idea in the face of recent efforts to dethrone it by a cadre of revisionist environmentalists, arguing for its necessity in arresting societal drift toward an arrogant humanism.

In his contribution, another of us (Ben) reflects on the enduring value of the classic wilderness aesthetic in a "post-preservationist" era, profiling the life and work of one of its most influential, though often misunderstood, artists. And yet as Kathleen Dean Moore writes, appeals to the traditional beauty and power of wilderness and wildness may in the end not be enough. To make good on biologist E. O. Wilson's striking proposal to place half the world under protection, Moore suggests we'll need to embrace not just the pristine wild, but also "feral" lands: places once "thoroughly trampled and trammeled," but that are evolving now outside of significant human influence and control.

The last section ("Knowing Nature in the Human Age") emphasizes the vital role of natural history and the importance of direct experience in keeping us rooted to nature (both wild and not), even in—perhaps especially in—the midst of significant ecological, social, and technological change. "We are what we pay attention to," Tom Fleischner argues, and we will only be able to keep a love of nature alive by choosing to turn away from the human and technological toward the natural and ecological. Martha Crump similarly writes how close observation of nature, when paired with a deeper emotional connection to it, can have a transformative impact, especially on children who haven't been fully captured by the tempos and toys of the digital age. And yet, as Susan Clayton and Bill Adams point out (respectively), technologically mediated ways of experiencing the wild, from virtual nature to video games, may not always be anathema to the ecological connections and the natural empathy Crump and Fleischner commend. Digital devices are often caricatured as the bane of conservation and wilderness ethics, but both Clayton and Adams illustrate how they can teach us important ecological lessons and values, perhaps even how to find "heart" in the digital wild. That's especially true if such technology can connect a demographically diverse audience to shared environmental experiences that don't require exclusive access to a shrinking number of remote wildlands.

It is an idea that's powerfully exemplified by Christopher Schell in his discussion of the imaginative and critical resources of Afrofuturism and Black joy, especially Schell's vision for how these expressions and meanings can inform a more equitable and vibrant social and ecological future. Indeed, as Joel Berger and Gary Nabhan both demonstrate, it may in the end be the human dramas—which in their essays include those of wildlife caretakers, curious naturalists, and the denizens of harsh ice and desert environs—that will inspire the most hope. They remind us that stories about nature are often most compelling when they are also stories about people, and about the human and ecological bonds that tether us together.

Finally, Harry Greene, in the book's afterword, closes out the discussion by advocating an ecological and aesthetic vision of the wild in which we're respectful participants rather than passive spectators. Most of our authors, including the two of us, have been influenced by Harry's life and work, which provides the personal, intellectual, and literary inspiration—and heart—of *The Heart of the Wild*.

———

Thomas Cole produced another noteworthy painting the same year as his *The Course of Empire* series, a large, panoramic work depicting the Connecticut River Valley in Western Massachusetts. It was a dramatically bifurcated image: The left side of the painting represented the wilderness with the depiction of characteristically rugged terrain, ominous storm clouds, and battered trees. A cleared valley with farms and groves occupies the right side of the canvas, highlighted by the river turning back on

FIGURE 4. *Top,* Thomas Cole, *View from Mount Holyoke, Northampton, Massachusetts, after a Thunderstorm—The Oxbow,* 1836. Oil on canvas, Metropolitan Museum of Art. Source: Wikimedia Commons. *Bottom,* detail from the bottom of the painting, showing the artist at his easel.

itself—a feature that drew tourists to the area during Cole's time. Titled *View from Mount Holyoke, Northampton, Massachusetts, After a Thunderstorm,* the painting is commonly known as *The Oxbow* (fig. 4).

One reading of Cole's painting is that he's making a visual argument for the harmonious blend of the wild and civilized on the American landscape. That is, the pastoral scene on the right side of the canvas is the golden mean, the "middle landscape" that contained the best of both worlds—wilderness and civilization—while avoiding the excesses of each. It's a reading that fits with Cole's often personal response to the raw wilderness, which could be complicated. Although he was enamored of the wildlands of his adopted home, like many nature preservationists (from Henry David Thoreau to Edward Abbey) he also felt pulled by many of the comforts, achievements, and above all, the companionship offered by society and settled life.

And yet there's another interpretation of *The Oxbow*, one that views it as less conciliatory and more challenging. Note that Cole painted himself in the scene, appearing as a very small figure in the bottom middle of the image hard at work on his easel. He's facing the viewer, and perhaps confronting us as well with a question: What future will we choose? In this reading, the shape of the river becomes a literal question mark on the land.

Nearly two centuries later Cole's question still confronts us, albeit reframed and modified for our own time. Can we keep a love of nature and the wild alive as we push deeper into this human age? Can we balance a respect for our valued environmental and conservation traditions while also recognizing the need for change—and the often jarringly different contexts presented by today's and tomorrow's ecologies, values, and peoples? What really resides at the heart of the wild, and what does this mean in an increasingly human-occupied and technophilic age? We hope the essays that follow can help us think through these barbed but vital questions as we continue to move, even if only metaphorically, across Cole's canvases.

PART I

Conservation's Shifting Ground

1

There Goes a Badger

Emma Marris

I WALKED alone on a concrete path in my neighborhood. After a while, I noticed that the compacted dirt and sparse weeds on either side had given way to lush vegetation. The sound of cars on the parkway had suddenly been augmented by the busy sounds of a flock of finches. I paused. Closer examination revealed that a pipe culvert ran under the path. On each side was a tiny spontaneous urban ecosystem taking advantage of the moisture, composed of teasel, cattails, millet, sunflowers, and asters, among others.

Water was running through the culvert. I knew roughly where it was going. It would pass under the parkway and find its way to the nearby irrigation canal, which would go on to deliver water to farmers, ranchers, and a National Wildlife Refuge south of town. But I wasn't sure where the water came from. On a whim, I decided to find out.

I stepped off the concrete path and began following the narrow band of water upstream, walking parallel to a band of cattails that indicated its position. After a while, I found an unofficial bridge over my urban freshet, a single partially rotted plank laid by a person or persons unknown. I found myself unexpectedly touched by the way that human animals had spontaneously adapted to and accommodated this unnamed rivulet.

No city official or Google Map can completely describe or comprehend the ecology of a city. A map might tell you the name of the parkway, the location of the concrete path. The culvert itself is possibly recorded on some specialized map of such infrastructure. And the ponderosa pines planted along the path, babies still, swaddled in their polyethylene tree tubes, were raised in a greenhouse and planted by some official entity— either the city or perhaps a volunteer club. But the cattails and the mullein and the duckweed I passed were never immortalized in any nursery inventory or bought or sold. The plank bridge was not listed on any map, not recorded on any survey.

The bridge, like the cattails, was outside of capitalism, outside of civic bureaucracy, outside of central planning. Both were produced by ecological forces. In the case of the cattails, their wind-spread seed found a cozy niche in the moist verges of the creeklet. In the case of the bridge, unknown humans used their signature tool-making skills to solve a problem of movement through their territory. The key to seeing the bridge as an instance of wildness is to remember that humans are animals too. Our propensity to shape ecosystems or, let's be frank—to raze them to the ground and replace them with alternative ecosystems—can be seen as "niche construction" as well as "environmental destruction." The two perspectives aren't even mutually exclusive.

In 1927, Charles Elton wrote, "When an ecologist says 'there goes a badger' he should include in his thoughts some definite idea of the animal's place in the community to which it belongs, just as if he had said 'there goes the vicar.'" Today, we struggle to remember to do the opposite: to remember that the vicar, too, is an ecological agent and subject, just like the badger.

I crossed the bridge, pausing midspan to watch water striders create perfect circles on the face of the waters. Then I continued upstream on the other side, feeling obscurely as if I had passed into a less predictable realm, a kind of backlot Narnia.

It was October 2020, the height of the pandemic, and I had been cooped up with my nuclear family in our house for months. My husband and kids were starting to feel like physical extensions of my own body. With every breath I inhaled their exhalations. I craved escape, solitude, the company of other species. And although I was less than a mile from home, ambling through an undeveloped slice of interstitial urban space, I began to feel the way I feel when I hike alone in a federally designated wilderness: I felt the sensation of cool water bathing my overheated, itchy, addled brain. I felt alone but also surrounded by multitudes, melted into the biomass. I felt like an animal.

I approached a small cluster of willows, which huddled around the creeklet, growing low and wide. As I stepped into their shade, I flushed a covey of California quail, their wingbeats and burbling chicky-chucky calls sounding as liquid as the creek itself. I sat in the shade of the willows for a moment and listened to blackbirds and flickers converse. Ants encountering my shoes scampered over my sneakers.

Appreciating urban wildness does not mean abjuring the pleasures of large, remote, protected areas. I have argued elsewhere that some of those areas are managed to the extent that they may be less "wild" than the willow grove where I rested. But I have also argued that wildness is not intrinsically valuable. Many humans *like* wildness, but it would not be valuable if there were no humans around to value it. Indeed, it would not exist if there were no humans around to value it, since it is typically defined as a lack of human influence or minimal levels of human influence. Without humans, the category would make no sense.

The category is, frankly, already quite rickety. So much of what white people are wont to call wild are places that were managed at various intensities by Indigenous peoples for millennia. Yet they were seen as "wilderness" because the orchards weren't planted in straight lines and because the fire-maintained meadows weren't enclosed, and because the wildlife managers and farmers were nearly all dead from smallpox and other plagues, and because the white people in America and the culture they created were incapable of seeing Native people as human beings with agency.

I think the value in the nonhuman world, and especially in places and organisms that are not being actively managed by humans, is actually in their many individual selves and in the complex flow of energy, matter, and information between them. Because of a history of human destruction of both these organisms and their interconnections, we have absorbed the idea that the human touch always destroys what is beautiful and good in this living web. For much of its run, Western conservation has focused on removing humans from the picture, from snipping us out of the web so we will stop destroying it with our thrashings. But humans can also be woven into the web without destroying it. Indeed, this is the only way we can preserve what we love in the nonhuman world, given that there are so many of us. We may want to withdraw from "nature" in shame but there is no place to which we can withdraw. We are all here on Earth together, quail and willows and humans, like strangers packed into the same ballroom. There are no exits, or if there are, they lead to cold space. Our only option is to learn to dance.

I continued upstream, following the water through another culvert pipe up to a dusty, hilly empty lot behind a retirement home. Here I found lots of intriguing evidence of nonhuman animal activity: shallow holes, clear paths through the dried grass, and mysterious and surprisingly prodigious piles of dung. I peered into the end of the culvert pipe. What beast might be sleeping off the heat of the day inside?

I followed a hint of white in the pale yellow dried grass and found a tiny skull, perhaps prey of the creature making all the massive scat piles. I traced its long front teeth, the bare bone of its jaw between those incisors and a cluster of slender molars. I felt the delicate zygomatic arch surrounding the empty eye socket. Later, I would identify it as a muskrat skull (fig. 5). Here then was the last remnant of a wild urban resident, an animal with no fixed address, no social security number, no pet license or GPS tracking collar—a truly anonymous creature who likely lived its entire life within city limits.

Holding the skull carefully, I followed the dwindling creek further up, under some trees and along the edge of the human-managed landscaping around the retirement home. The flow had reduced to a trickle, and now, as I scrambled up a slope spangled with trash—photodegraded tinsel, an empty can of Copenhagen chew—it had become no more than a wet spot in the dirt. Following this, I came to the headwaters of my nameless waterway, a curb cut at the lowest corner of a large, sloped

FIGURE 5. The muskrat skull. Photo by the author.

parking lot belonging to a community health center. Here, the accumulated runoff from some 10,000 square feet or so of impermeable pavement collected and flowed on downward through the city, following no
other laws but those of physics.

I walked out onto the parking lot and saw a hand-painted sign outside
a hastily erected shed. It read, "COVID-19 TESTING: DRIVE THRU
ONLY." The skull's meaning shifted in my hand, from natural history
curio to memento mori. My escape from the pandemic, from the
human-centered realms of the city, was over.

And yet the untold billions of copies of SARS-CoV-2—the virus that
causes COVID-19—are also "wild." They are part of the earth's web of
life and quasi-life. They are not under our control. They remind me why
so many people throughout history thought of "wilderness" as a bad,
scary place. Here, on this concrete slab, a landscape designed to exclude
all other species, to smother life and cauterize the soil, I found one of

our desperate rearguard actions against a beast with just fifteen genes that has killed more than four million of us.

If "wilderness" describes a whole landscape or ecosystem that is uninfluenced by humans, then "Wilderness" is mostly a colonial myth. If it does exist anywhere, I maintain that it is only valuable insofar as some humans value it. I don't use the word myself, except to debunk it. The word "wild" is more complicated. Wild has been used to describe ecosystems, but also to describe individual organisms. As it applies to plants and animals, it can carry various meanings, from "genetically un-domesticated" to "behaviorally autonomous." There's a vast difference between human influence and human control. The little stream that I had followed would not have existed without the parking lot but its inhabitants, from the flickers to the sunflowers to the muskrat whose skull rested in my hand, all lived autonomous lives on their own terms. In our world, yes, but not under our thumbs.

In my book, *Wild Souls*, I argue that one kind of wildness is probably morally valuable: the autonomy of individual sentient animals who want to make their own choices. Which animals, precisely, this would include is an open question. But many animals become clearly upset or despondent if they cannot do what they want—if they are prevented from migrating or mothering or hunting. Insofar as being able to make their own choices is part of their flourishing, their behavioral "wildness" is valuable to them. But this applies only to sentient creatures who make choices. Not all "wild" things—in the sense of things beyond human control—are equally valuable. Few among us would hesitate to obliterate every coronavirus on the planet if given the chance.

The value of "nature" as we generally think of it primarily rests, as I have argued, in an ecosystem's many individual selves and in the complex flow of energy, matter, and information between them. Autonomy may be valuable to some of those individual selves. But beyond that, I do not think one even needs the concept of wilderness or wildness to create a robust environmental ethic. And yet, I do admit that I found my walk off the paved trail all the more charming because the plants and animals and waterway I found there were not planned by humans or even completely known by humans. You could call this an urban wilderness—but you could also call it a multispecies city—a surprising multispecies city. All of us urbanites can find such surprises, such non-human spaces, if we look.

A few weeks later I heard that a badger had been spotted in the neighborhood, and this remains my best guess as to the denizen of the culvert out back of the retirement home. A badger, I think, would eat a muskrat. There, apparently, goes a badger.

2

Embracing the Cane Toad

Richard Shine

Compromise does not mean cowardice.
—JOHN F. KENNEDY, *PROFILES IN COURAGE*

Why We Should Try to See the Glass as Half Full Not Half Empty

Few relationships are as intimate as that between naturalists and their study sites—and especially, their study organisms. So, the loss is visceral as those special places are degraded and defiled, and those magical animals or plants cease to exist. And the scale of that loss is horrific. From childhood, I have adored frogs, and the predatory snakes that depend on them—and over the course of my career I have seen scores of amphibian species vanish into extinction, and with them the thriving populations of elegant serpents that I took for granted during my early-career studies. It hurts.

How should I react to that catastrophe? By railing against the forces that caused it? But those causes are diverse. An invasive fungus that is fatal to frogs. The greed of farmers who stole the river's water to irrigate their crops. The failure of political leaders to recognize and respond to climate change. The swelling abundance of my own species, converting wild places into suburbia or worse. Yes, that outrage should be given voice, and to do so is cathartic. But it's not enough, and it can't be our main response.

Imagine, if you please, that you are the only surviving child of profligate parents. In your early years, the walls of your family home were adorned with priceless artworks, but those were destroyed through decades of neglect and misadventure. All that are left are tiny fragments of that former glory. What do you say to your own children, to encourage them to preserve those last vestiges of the splendor that you recall so clearly, and whose loss weighs so heavily on your private thoughts? Complain about the loss? Blame your parents for their thoughtlessness? For their failure to preserve a unique heritage? Nothing can restore the full splendor of that art. All you can do is try to motivate your own children to preserve what is left, and perhaps, create something new.

Giving up is not an option; there is too much at stake. Although much has been lost, portions of that heritage remain. And your children's actions will determine whether those remnants also disappear into history, or whether those fragments can be maintained. The key to nurturing that conservation effort across time—long after you are gone—will be the attitude that you encourage in the next generation. If they find joy in those objects, they will try to preserve them—and perhaps even help to create new "ecological art" by methods such as rewilding. And if they do not find that joy, the myriad pressures of modern life will see the fragments ignored and, eventually, lost.

So, there is no point extolling the virtues of that long-gone vista—the collection of art on the wall. Or of describing the intricate details of a masterpiece that has now crumbled. Instead, point out the beauty of what has survived. To you, it may seem a squalid remnant of something much vaster and more awe-inspiring. But your children never saw that original piece of genius. Their baseline has shifted. To them, that small surviving picture is the only thing that matters. And if they value it— rather than seeing it as evidence of a catastrophe—they may be willing to preserve it.

What does this mean for a naturalist? The works of art that we prize are natural ecosystems and the living creatures that fill them. Enormously more beautiful and important than any artifact created by human hands.

And for a generation of naturalists reaching retirement, it is difficult to look beyond the horrific loss of once-magnificent entities.

There are still a few places in the world where you can capture the feeling of human insignificance in a breathtakingly large area that has escaped anthropogenic trashing. But such places are increasingly few and far between. The rainforests of the Amazon, the deserts of Central Australia, the icebound lands of the Antarctic. But almost everywhere else, the land bears the imprint of humans. And even in the remote places, disturbance is growing. Electricity lines run along the bank of the Amazon. Tourism and mining are changing the desert. The ocean is awash with plastic contaminants. And a warming climate is removing Antarctica's icy covering. Increasingly, the idea of a pristine wilderness is as real and alive as a dodo or a thylacine. As naturalists, we need to accept that reality. By and large, "natural" ecosystems—functional entities unaffected by people—are a thing of the past. They will never return, unless the human race succeeds in exterminating itself. So, we need to move through the stages of grief—past denial, anger, bargaining, and depression—to the final phase of acceptance. If we truly want to save what is left—even if it is but a tawdry remnant of what we once knew—we need to treasure it for its own sake instead of seeing it as a reminder of what we have lost.

Shifting Baselines

By the measures we usually employ when evaluating the ecological attributes of a species, *Homo sapiens* is a towering success. We dominate the planet. We are the ultimate ecological engineers, modifying habitats in ways that exterminate many taxa but enable a few to thrive. And the main driver of that success has not been genetic adaptation to local conditions but, instead, our enormous behavioral plasticity. The genetic differences between local races of people are small, and yet phenotypes constructed by that human genome thrive in almost every climate and habitat type on Earth.

That flexibility takes many forms, but one vital aspect is the way that we modify our expectations based on recent experiences. One well-supported concept of brain function in humans emphasizes the role of forecasting; we constantly predict what is likely to happen, and we compare that prediction to reality as it unfolds. If the prediction succeeds, we ignore it. But when the prediction fails, it alerts us to a potential problem— something is happening that we wouldn't expect, based on previous experience—and thus, it warrants close attention. Our brain focuses its attention on that aspect. If we throw a ball in the air, we expect it to come down again, and can foretell its trajectory. If that happens, we barely notice. But if the ball keeps moving upward, we are surprised, and we pay attention.

Critically, though, the prediction is not static. It shifts with recent experience. If every ball we throw into the air disappears from sight, it no longer surprises us. We may not understand why it's happening, but after a while we expect that outcome. And we modify our behavior accordingly, no longer positioning ourselves to catch the returning ball.

Although humans are especially adept at shifting expectations based on experience, it's by no means unique to our species. During fieldwork in Manitoba with garter snakes (*Thamnophis sirtalis parietalis*), my colleague Bob Mason and I noticed that males consistently prefer to court large females rather than small females. Bigger females produce more offspring, presumably enhancing the fitness benefits of a copulation for the lucky male. Hence, a preference for courting larger rather than smaller females may pay off in evolutionary terms. But what if large females are scarce? A male that refrained from courting smaller females would lose the only opportunities available. So, we predicted, selection should favor flexibility: a male snake's preferences should depend on his recent experiences. And sure enough, removing all of the larger females around the communal den induced a rapid shift in sexual targeting. Males began to court small females that they would normally have ignored (fig. 6).

Humans take it one step further than garter snakes, though. We constantly take in information, compare reality to expectation, and are quick to change our expectations (and thus, our behavior) if the mismatch with reality is too great. Of course, there are limits to that flexibility; people can stubbornly hold on to expectations in the face of overwhelming evidence. Every week, I expect my favorite football team

FIGURE 6. Red-sided garter snakes (*Thamnophis sirtalis*) at a communal den in Manitoba, with a small male courting a larger female amid the chaos of thousands of other amorous males. Our research showed that males prefer to court larger rather than smaller females, so long as large females are available—but if they are not, males switch their attention to smaller targets. Photograph by Tracy Langkilde.

to win its next game, and almost every week, they lose. But that constant mismatch doesn't change my fervent belief that victory is just around the corner. Likewise, rusted-on supporters of right-wing political pundits maintain their support even as the world warms around them and a pandemic slaughters their like-minded unvaccinated compatriots. But in many other issues, humans are far more flexible and more willing to change expectations. And as the natural world has become degraded and an increasingly urbanized population loses personal contact with nonurban habitats, our expectations about nature have changed dramatically. Few baselines have shifted as far and as fast as our perceptions of wilderness.

In their writings, many environmentalists deplore these shifting baselines. Longing for a bygone era, elderly naturalists claim that we will never be able to preserve our natural heritage if we accept constantly diminishing prospects. If we treat degraded habitats as equal in value to the pristine ecosystems that they have replaced. But if returning to some long-lost scenario is impossible, then there is virtue in accepting what you still retain. Further diminishing its value, by comparing it to what came before, achieves little. Perhaps shifting baselines are our ally, not our enemy, in the fight to preserve wilderness. It helps us to see the glass as half full not half empty.

Degraded Expectations

Like many of the naturalists who contribute to debates about nature conservation and wilderness—including some of the authors in this book—I had my first experience of wilderness several decades ago. But the people who will decide the future of wilderness—the youngsters, just finishing high school and college—have grown up in a very different world.

Most naturalists vividly remember their first intimate exposure to wild places. Where I spent my boyhood in southeastern Australia, the first reptiles on offer to me were small lizards in suburban gardens. But undeveloped valleys surrounded by sandstone cliffs were just thirty minutes' walk away, and a snake-obsessed boy could wander to his heart's content in bushland with little overt signs of human disturbance. It was a society that did not yet prohibit the collection and retention of wild animals as pets, and where children were allowed to take risks. Where the presence of pedophile strangers was unsuspected, and children meandered without supervision. Television was in its infancy, and video games and social media not yet available; and thus, there were few distractions to keep a young wildlife enthusiast at home.

But that world was changing. At the same time as peri-urban bushland was being degraded by "development," the political winds were shifting. Outdoor activities that caused stress or harm to individual animals were frowned upon and eventually outlawed, under the influence of animal-welfare-inspired ideologies. Increasingly, people questioned the morality of hunting and fishing. And the idea of capturing a wild animal and keeping it in a cage was viewed as barbaric. The end result, inevitably, was a schism between people and wildlife. At the same time as wild places and the wild animals were disappearing through destruction of habitats, society was disdaining the consumptive use of wildlife.

It is difficult to overstate the magnitude of that shift. In leafy Sydney suburbs during the 1950s and 1960s, animals were everywhere. Backyard chicken coops were common, and children grew up with a hands-on interaction with living creatures. We saw where eggs came from, and we saw how a live chicken was turned into Sunday roast. We fished for eels in the local creek, and cooked and ate them over a campfire. We slept out beside the river, marveled at snakes and kangaroos that came down to drink, and eventually graduated to roaming the bush shooting rabbits and foxes. We set tame ferrets down rabbit holes, and waited expectantly to capture the terrified rabbits that came hurtling out.

And before the relentless expansion of suburbia, many homes were close enough to remnant bushland that the wildlife came to us rather than our having to go to them. I caught bandicoots at night on the back lawn to feed to the prized pythons I had captured in adjacent woodland (fig. 7). My father's hopes for his roses were dashed, year after year, when possums consumed the flowers before they could open. And if you included the diverse menagerie of lizards and snakes in makeshift cages in the backyard and garage, I interacted with dozens of living creatures every day.

That world no longer exists in most of Australia. Children still encounter the idea of animals, mostly in a digital form on a screen, and have direct contact with live beasts in highly structured circumstances like classrooms and zoos. But apart from dogs, cats, and goldfish, few children experience a real intimacy with animals—and especially, with the wildlife that once lived all around their homes.

A concurrent shift has occurred in the training of biologists. For example, dissections of real animals are being phased out for the cheaper and less confronting alternative of virtual dissections. More importantly, field-based courses have largely disappeared from the curriculum in many institutions. Although popular with students, such courses are expensive, they risk legal liability for injury or mishap, and they require

FIGURE 7. The author as a youngster, holding a recently captured and (slightly) venomous whip snake (*Demansia psammophis*). Children today are discouraged from such hands-on interactions with wild creatures. Photo by the author.

more and more infrastructural support as students become less and less experienced with life outdoors. Additionally, many institutional ethics committees refuse permission for students to interact directly with wild animals, on the basis of animal-welfare considerations. On one recent course I taught, the ethics committee stipulated that a given animal could be physically handled by no more than two students, to minimize stress. Even the small subset of people who are being trained as wildlife biologists are forbidden to touch a wild animal.

In short, we have seen a growing division between humans and other species. The concept of "wildlife" is increasingly about images on a television screen rather than personal contact. Australian children know more about lions and elephants than about bandicoots and possums. And although the quality of that digital imaging has improved dramatically, it is no substitute for personal encounters. Wildlife becomes a vague concept, equivalent to the cartoon characters that crowd the TV screen. And if children do retain an affection for wildlife, they believe that it is a sacred but distant assemblage of endangered species; and that the only responsible attitude is to avoid contact with such desperately imperiled creatures. To a professional biologist, "conservation" implies management, often entailing pragmatic, active interventions to maintain a species' viability in the face of looming threats. But that same word, "conservation," has taken on a different meaning for our children. "Conservation" is simplified to "preservation" or more generally, to "avoidance." Driven by multiple societal influences ranging from animal-rights activism to spiritual tree-hugging, our children are taught that touching a wild animal is an act of vandalism.

How Can We Keep the Curiosity Alive?

A child who grows up without personal contact with animals is unlikely to be an advocate for wildlife conservation later in life. With the best of intentions, notably from the animal-welfare lobby, we have spawned a generation who see wilderness and wildlife as intellectual or quasi-spiritual issues, without the visceral response experienced by people in close contact with the wild. When political priorities are assigned, conservation has to compete with pandemics, terrorism, child protection, and the war on drugs. And wildlife conservation has become just one of a broad array of issues that concern the modern world, ranging from the love lives of celebrities to the tribulations of football teams. Unless wilderness resonates on a personal level, it will be lost amid those dazzling distractions.

What can we do to reverse these trends? To convince younger people to care—deeply, emotionally, and irrationally—about the natural world? To adopt a broader view of humanity and our place in the world, a view that incorporates the glory of the other species who are our fellow-travelers on Planet Earth? In the end, the answer will be about personal connections. Experiences that resonate, at some deep level, and shape who we are. And digital inputs via an electronic screen cannot replace hands-on encounters.

But if that conclusion is correct—if we need to bring children into physical contact with wildlife and with wilderness—the species involved, and the nature of that wilderness, will be very different to the situations that inspired an older generation. I had the extraordinarily good fortune to grow up in a world where I could spend hours every week in bushland, seeing no other people, and I could observe (and sometimes capture) giant lizards and deadly snakes. That experience set me on a path to becoming a professional naturalist. But those days are gone. The bushland has shrunk, the animals have disappeared or are highly protected, and most parents view the outside world as a place so hostile and dangerous that children cannot be left unaccompanied.

And so, we have to shift our baselines. We can't expose our children or our grandchildren to dramatic beasts in huge tracts of wilderness, but we *can* show them small lizards and frogs in the local park. Creatures and places that look humdrum to older eyes can be magical for people seeing this modest "wilderness" for the first time. The only "wild" landscapes and animals close to our children's homes are pedestrian, by comparison with the experiences of an older generation. The bandicoots and pythons are long gone from most suburbs of Sydney, but the gardens are home to many smaller creatures, and the scrubby edge of the local football field still contains fantastic organisms.

For an ecologist of the Old School, one of the perennial challenges is the fact that much of that suburban biota consists of species that should not be there. Despite the best efforts of community environmental groups, the native bushes and grasses have been replaced by invasive shrubs and pasture grasses, many of them spreading from gardens nearby. The small birds, so dependent on thickets to breed, died out as the bushland was cleared and feral cats intruded. All that we have left today are larger birds, opportunists that have benefited from the decay and homogenization of those original habitats. Once rare, the larger birds now dominate. And among the taxa that embrace habitat disturbance, a horrifying proportion are alien species rather than native taxa. Indian mynah birds throng the streets. Cane toads call from the edges of the local pond. A fox runs across the road as I drive home from work.

Like many ecologists, I have spent much of my professional life railing against the dangers of invasive species, quantifying the impact of those aliens on native ecosystems, and devising ways to control the continuing spread of the invaders. And an important part of that effort has been to educate the public about the dangers that invaders pose, and about how to recognize and exterminate them. I don't regret those efforts, and I treasure our modest successes. Given the choice, I would prefer an Australia full of quolls rather than cats. But increasingly, that choice isn't available. The habitats that support high densities of humans are tough places for other species to make a living, so the taxa that overcome those challenges comprise a motley crew of natives and invaders. This is the New Normal. And for my grandchildren, it is the Only Normal. They never saw those ecosystems before they were changed profoundly.

Confronted by that decimation of biodiversity, it is tempting to talk about ecosystems being trashed, and about the need to turn back the clock and restore these once-magical places. That's a wonderful aspirational goal, and I embrace it with all my heart. But at best, it's a long-term aim. At worst, it's a futile dream. So, the older generation of naturalists faces an existential challenge: we despise the modified ecosystems around us, but they are the only places in which our children can experience even a taste of "wilderness." How can we teach our children to love the wild, if the only wild places within reach are irretrievably broken?

Every generation experiences change, and struggles to accept the new realities that supplant cherished customs and situations. But the older naturalists among us are facing an upheaval on a vast scale. Should we embrace the degraded habitats and invasive species that surround our cities, as the only kind of "wilderness experience" available to our children's children? Pragmatically, we need to. We can still keep fighting the good fight, hoping to sway people toward prizing some aspects of an ecosystem more than others, and we can try our hardest to bring a younger generation into contact with the remaining truly wild places. But there are a lot of people, and very few places that function ecologically in the ways that they did a lifetime ago. Even if we *could* get hordes of those people to the wild places, it would be futile. With people everywhere, those places would no longer be wild.

So, I think, we need to be forgiving. We can't avoid an emotional reaction to ecological degradation, but we need to separate that reaction from our professional aims, like restoring habitat for native wildlife. A small patch of peri-urban bushland can be beautiful, even if most of its trees came from other continents. A cane toad (*Rhinella marina*) is a fantastic form of life that has overcome unthinkable challenges to spread successfully across the world. And by shifting our baselines—by seeing the concept of wildness embodied even in a modest urban park or an invasive toad—we have an opportunity to transmit our passions for the living world.

The Ubiquity of Disturbance

Embracing the willow tree and the cane toad as legitimate objects for veneration—as valid occupants of Australian ecosystems—might be a bridge too far for many environmentalists. But that shift in attitudes need not preclude our strenuous attempts to replace willows with eucalypts, or cane toads with native frogs. It would just mean that we are replacing something of value, with something of even more value. To acknowledge that all forms of life have positive aspects, rather than seeing some taxa (especially aliens) in a totally negative light. Of course, that approach eventually breaks down—few would extoll the virtues of ebola virus or COVID-19—but broadly, we can look for positives to transmit to the generations who follow us. The important aim is that they should venerate wilderness and wildlife, not that they should venerate the same notions of wilderness and wildlife that drove their parents' passions.

On a personal level, I am comfortable with environmental disturbance. I enjoy near-pristine wilderness—it is an aesthetic delight—but most of my long career in fieldwork has been in habitats that have been grossly disturbed in one way or another. I worked in those places because they gave me access to abundant study organisms—in my case, snakes—and the disturbance was a critical positive for me. It enabled snakes to reach high densities, or to be easily observable, or it made it logistically feasible for me to visit a site and conduct research.

Sometimes the disturbance was subtle, not obvious from casual inspection. For example, the massive communal dens of garter snakes (*Thamnophis sirtalis*) in Manitoba support tens of thousands of snakes, a paradise for researchers. The absurdly high biomass of snakes is possible only because of food resources, and especially the profligate abundance of earthworms—most of which are invasive species. We may never know how many snakes inhabited those dens before alien earthworms penetrated into the Canadian prairies. In other cases, the disturbances are wrought by natural processes, but maintained by human activities. For example, the amazing densities of pit vipers (*Gloydius shedaoensis*) on the tiny Chinese island of Shedao depend not only on an annual food subsidy—the wholescale migration of passerine birds between Siberia and Southeast Asia—but also on the availability of ambush sites from which a snake can seize a bird. Thick forest offers few such open sites, but hurricanes rip holes in the canopy, and people maintain those clearings. Without those disturbances, Shedao could not maintain a density of one snake per square meter.

Most of the places where I have studied snakes exhibit far more obvious anthropogenic vandalism. By constructing a dam wall across a floodplain in tropical Australia, rice growers created a habitat mosaic that rendered snakes both abundant and accessible. By removing large predatory fish, and providing a constant inflow of freshwater, the inhabitants of the city of Noumea created a coral reef habitat ideal for small fish and the sea snakes that feed upon their eggs. By modifying the environment, people destroyed much of the beauty of those sites, and they removed that evocative feeling of being in the midst of nature. To a romantic soul, those sites are trainwrecks. But to a pragmatic researcher, looking for opportunities to unravel the complex secrets of poorly known animals, they were a godsend.

And so, should we maintain the rage, and eschew all value to modified ecosystems? Put our efforts into saving the largest patches of near-pristine wilderness? Or should we accept our losses and work to enhance biodiversity values even in highly degraded places as well—because these are the places that most children will experience? And are we prepared to look pragmatically at what can be achieved, and accord real value and meaning to an alien species—even if it has been ecologically devastating to the native taxa that it has replaced?

Tragic as that conclusion is, at so many levels, we need to move with the times. Those of us who knew a better world may never be able to wholeheartedly embrace the degraded ecosystems—the cane toad calling beside the trash pile—but we need to teach our children that even a quasi-wilderness is better than no wilderness at all.

3

Invasive Species in the Anthropocene, or Learning to Love the Dingo

Jonathan B. Losos

THE DINGO is an unlikely hero. Long and lean, with a large head and powerful jaws, it's not hard to imagine these tawny canines winning best of show at Westminster. But dingoes are not a human-sculpted dog breed; ever since people brought them to Australia 3,500 years ago, dingoes have been reverting to a wild, wolf-like animal, living in packs and hunting Australia's finest.

And that's why this handsome creature has historically received little love. Ranchers despise them because of their tendency not only to kill cattle and sheep, but to kill them in great numbers, leaving many uneaten.

Environmentalists, too, have long considered the dingo a villain. A dog gone wild, dingoes were held responsible for terrorizing the native fauna, running down kangaroos and emus, and wiping out many native species. Dingoes seemed like a textbook case of why introduced species have long been considered verboten in environmental circles; plopped into a land in which the native species have no evolutionary experience—and thus no defenses—such species often wreak havoc on native ecosystems.

FIGURE 8. Dingo puppies at den, Ormiston Gorge, Northern Territory, Australia. Photograph by Mike Gillam.

And then there was the sad case of Lindy Chamberlain. Here in the United States, the dingo is probably best known as the laugh line from a *Seinfeld* sketch, when Elaine, in a terrible Australian accent, exclaims, "Maybe the dingo ate your baby." The quip isn't so funny when you learn that it's a reference to a woman who served three years in jail for murdering her two-month-old until her conviction was overturned and her explanation, that a dingo had snatched her baby from a campsite, was accepted as true. In fact, since the killing of baby Azaria in 1980, there have been many dingo attacks on humans, some of them fatal.

In sum, the dingo was a textbook example of the horrors of nonindigenous species: wiping out native species, causing massive economic damage, even killing people. Back when I was growing up, the dingo had few fans. Public policy in Australia was to try to get rid of them by any means possible.

Times have changed, both for the dingo and for introduced species in general. Many people, conservationists among them, have come to realize that non-native species are not inherently objectional and that many inevitably will occur in, even be desirable members of, future ecosystems.

———

The geographic distribution of species exemplifies the old adage that the only thing that is constant is change. Over the billions of years of life's history, species have constantly been on the move. When the climate has heated up, species have expanded northward and to higher elevations; when it has chilled, they've moved in the reverse direction. Occasionally, one or a few individuals have taken a great leap, crossing an ocean or a desert, establishing a new outpost. Earth history, too, has affected species distributions, colliding continents or lowered sea levels permitting expansion into new areas.

This arrival of species into areas where they didn't previously occur has sometimes spelled disaster for local residents. Several million years ago, the rise of the Isthmus of Panama allowed predators from North America to venture into South America with devastating effects on that previously isolated continent's mammalian herbivores. Similar out-comes on both land and sea have resulted from other biological inva-sions from when two formerly separated areas have been connected by geological events.

On the other hand, arrival in a new place can also provoke evolution-ary exuberance as new species take advantage of available ecological opportunities. Darwin's finches in the Galápagos are the textbook ex-ample, but there are many others: lemurs in Madagascar, cichlid fish in the East African Great Lakes, and lizards on Caribbean islands, just to name a few.

Not only is movement a natural process, but it's one that is going on today. Many species are currently on the move as a result of climate change and human modification of the environment, especially the removal of predators and competitors. Opossums, for example, were found only in the southeastern United States in precolonial times, but today have waddled as far north as southern Ontario. The nine-banded armadillo crossed the Rio Grande in the middle of the nineteenth century and now is well into Missouri, Kansas, and North Carolina. And then there's the coyote, a species that in 1900 was limited to the western half of the United States, southwestern Canada, and Mexico. Now, thanks to their own wiliness (yes, the name of Bugs Bunny's nemesis was well chosen) and the elimination of larger predators, coyotes cover the entirety of the United States, much of Canada and, having crossed the Panama Canal, will soon enter South America.

The take-home message here is that species movement is a natural process that has been going on throughout Earth's history and that the effects of such movements have helped to shape the world's biodiversity.

———

The big concern, of course, is not species who find their way to new lands on their own, but rather those that are brought to new places by humans. Conservationists have traditionally been opposed to all intro-duced species, but recent years have led to a backlash to this view. One thread of this recoil is exemplified by a conversation I recently had with a non-biologist friend. I told him that the domestic cat had been named as one of the world's 100 worst invasive species, the reason being that cats introduced to formerly feline-free places, like Australia and oceanic islands, had slaughtered the naïve local fauna. Rather than expressing alarm at the demise of so many species endemic to these places, my friend surprised me by his sanguine response, stating that this was nature's way through the eons: out with the old, in with the new.

Although my friend has a point, he overlooked one important detail: the rate at which species are becoming established in new places, thanks to human transport, is vastly greater than in times past, from tens of thousands to a million times higher in some places. In Hawaii, for example, species have arrived recently at a rate 700,000 times greater than during prehistoric times. So, what's going on today is by no means natural, if by "natural" you mean how the world was before we came along.

Just as in the past, some of these human introductions, deliberate or not, can ravage local ecosystems. For example, the brown tree snake, native to Australia, New Guinea, and Indonesia, arrived on Guam in shipping, then ate most of the island's native forest avifauna (nine of eleven species) into extinction. Similarly, the cane toad, brought to Australia in an ill-advised attempt to control agricultural pests, had a devastating effect on naïve native predators, which did not know better and succumbed to the poison in the toads' neck glands. Many other examples—plants as well as animals—attest to the negative effects introduced species can have on native biodiversity.

For this reason, introduced species have been one of the major concerns of conservationists. The late, great biologist Edward O. Wilson coined the acronym "HIPPO" to refer to the five major causes of species endangerment: habitat loss, invasive species, pollution, population (human), and overharvesting. I grew up thinking all of these had to be stopped. Introduced species are bad. Period.

In recent years, a more nuanced view has taken hold. Most introduced species probably have little effect on the ecosystems to which they're introduced. I've seen this myself with the Caribbean lizard species I study. Many species have been translocated by people to other places, but most of these introduced populations, though firmly established, do not expand beyond human habitations into natural habitats and thus have minimal effect on the native fauna and flora. Other introduced species do not grow to large population sizes. These nonimpactful species aren't much of a problem and aren't worth using scarce conservation funds to eradicate—better to save the dollars to eliminate the species that are truly threatening native species and ecosystems. This approach has now become standard. Of course, it can be hard to predict, a priori, which species, if introduced, will become invasive (the latter term reserved for those introduced species that have detrimental impacts), so it's still important to minimize the number of new introductions.

———

A new perspective has been advanced that fundamentally disagrees with the idea that introduced species are a problem. This view is multilayered. On a philosophical level, it equates opposition to introduced species ("nativism") with xenophobia directed at human immigrants. This is a false equivalence. The former is a view based on the well-documented negative effects of non-native species on the species native to an area, whereas the latter is simply human prejudice. How conservationists should treat the introduction of non-native species is debatable, but it is not helpful to draw inapt comparisons to political arguments. Moreover, as Dan Simberloff has noted, "There is nothing necessarily xenophobic about a desire to preserve traditional biocultures, which is motivated by the same general principles supporting the right of human societies to maintain their cultural distinctness."

A more empirical point relies on the observation that biodiversity increases as a result of human introductions. That is, the number of new species in an area resulting from introductions is greater than the species that have gone extinct, leading to a net increase in local species richness. Some have gone so far as to conclude that, in many cases, invasive species are good for ecosystems and the services they provide.

This view highlights the question: What is it we are trying to conserve? Thanks to human introductions, local biodiversity may be increasing despite ongoing extinctions. But what we are doing is replacing the unique species endemic to particular areas with the same set of species—those able to thrive in human-altered environments—everywhere, a phenomenon referred to as the "McDonaldization" of the environment by some and a "planet of weeds" by others (where weeds refer to species that thrive in human-disrupted habitats, like dandelions, pigeons, and rats). If nothing else, such global homogenization would lead to decreased numbers of species globally, even if species richness is enhanced at the local level—the great diversity of species differing from one place to the next will be replaced by the same species everywhere.

Even though the views just mentioned are misguided, those advocating for the beneficial aspects of introduced species do make some good points.

For example, are species moved from one place to another by humans labeled forever as introduced and non-native? How long does a species have to exist in a place before it earns its passport and is considered native?

Dingoes have been Down Under for nearly four thousand years. Isn't that long enough for them to be considered a legitimate part of the Australian fauna? Proponents of this view have pointed out that the dingo has evolved distinctive features, so much so that it might be considered its own subspecies or even a species, *Canis dingo*, distinct from the dog and the wolf (though others dispute its distinctiveness).

Another canine whose species-level distinctiveness is not in question, the Channel Island fox, presents a similar situation. This diminutive (and adorable) descendant of the gray fox occurs on most of the Channel Islands off the coast of southern California. Some evidence suggests that ancestral gray foxes were taken from the mainland to the northern Channel Islands by the Chumash people seven to nine thousand years ago; the alternative possibility is that somehow the ancestral foxes made their own way across the Santa Barbara Channel to the islands. Regardless, it is almost certain that the Chumash subsequently moved the foxes from the northern to the southern Channel Islands, which are too distant for the foxes to have floated across to on their own.

The fact that no one has accused the island fox of being a non-native member of at least the southern Channel Islands fauna suggests a rule of thumb: once an introduced species has evolved enough to be considered a different species, then it may be considered native to wherever it occurs regardless of how it got there.

Proponents of introduced species also suggest that sometimes there will be a positive conservation effect of their introduction. Some introduced species are threatened or endangered in their native range. For example, the Javan rusa, a type of deer threatened in its native range in Indonesia, has been introduced to many other places. Wouldn't it be better to have Javan rusa existing somewhere in the world rather than going extinct? Some have even proposed introducing endangered species such as rhinos to appropriate habitats in Australia or elsewhere as insurance populations should they go extinct in their native lands.

A related issue concerns how we treat feral populations of domesticated species. Such populations are doubly unnatural because not only are they not native to an area, but they also owe their entire existence to human intervention in the domestication process. Prima facie, one would think they don't belong out in the wild, particularly if they are having detrimental effects on the native fauna.

But if the dingo is so distinctive that it might be considered a distinct species, what difference does it make that it's a descendant of the dog? Indeed, many feral populations of domesticated species seem to have lost most or all vestiges of their domesticity.

And there's a further wrinkle. The wild ancestors of many domesticated species are now extinct. Consider the familiar barnyard cow and horse. Their wild ancestors are no more. Isn't it better to have wild horses and cattle that, as far as we know, are somewhat similar to their wild ancestors, than no such species at all? The dromedary was domesticated about four thousand years ago, and their wild ancestor is extinct. Yet, dromedaries released in Australia have spawned feral populations that are quite numerous. Are these populations an introduced pest, or might they be considered a conservation success story, or maybe both?

———

These are interesting points, but there's a much bigger reason why we might want to tolerate—even welcome—introduced species, at least in some circumstances. Much as we may not like it, the world is a very changed place from a hundred, a thousand, and ten thousand years ago. Ecosystems have been disrupted by the disappearance of many species, species that are gone (probably) forever. Is it better to have an ecosystem lacking some of its vital parts, or to replace those parts with similar species from elsewhere?

The dingo is, once again, a possible example. Before its arrival, the top predator in Australia was the thylacine, a marsupial predator strikingly similar to the gray wolf, though somewhat smaller. The dingo is probably responsible for the thylacine's disappearance from the Australian mainland (it persisted in dingo-free Tasmania until the early 1900s). There's no doubt that the dingo did a job on many moderate-size Australian mammals, many of which went extinct.

But they're gone. And the real problems in Australia today are two other introduced predators, the red fox and the domestic cat, which are having devastating effects on small vertebrates throughout Australia, species that are generally too small to be preyed on by dingoes.

We don't usually think of predators as prey, but in fact it happens all the time. Large predators the world over are renowned for killing, and sometimes eating, smaller ones. Lions kill leopards and hyenas, leopards kill cheetahs, wolves kill coyotes, which kill foxes, and so on.

Australia is no different. Dingoes suppress red fox populations and perhaps domestic cats, too (it's a little complicated because red foxes also have a negative effect on cats, probably more due to competition for resources rather than direct antagonism; hence the effect of dingoes on cats is complicated: dingoes relish a good meal of cat, but also, by killing foxes, have a positive effect on cat populations).

Of course, we'd rather have thylacines or even Tasmanian devils keeping foxes and cats in check. But until the thylacine is genetically resurrected (don't hold your breath), the dingo may be the best bet to restore Australian ecosystems by regulating the population size of middle-sized predators, as well as of large herbivores like kangaroos (kangaroos are very abundant in Australia and in a way similar to white-tailed deer in the United States; they are prone to population explosions that are bad both for kangaroos and the vegetation).

Whether the dingo is Australia's ecological savior or a damnable introduced species is still hotly debated in Australia, so let's consider a much less controversial example of ecosystem restoration by introduced species.

Large tortoises used to be common on islands throughout the world. In the Indian Ocean alone, there were at least eight different species on Madagascar, Mauritius, Réunion, Rodrigues, and the Seychelles. Now all but one of them—the Aldabra tortoise—are gone, wiped out by humans.

Indian Ocean islands, like islands in many parts of the world, have an odd fauna, the result of the somewhat random hodgepodge of species able to make their way across the ocean to become established. As a result, on many islands, tortoises were the largest herbivore, sometimes reaching more than five hundred pounds and attaining enormous population sizes (indeed, the total weight per square mile of Aldabra giant tortoises is substantially greater than that of all large mammals combined in African parks).

The removal of the dominant herbivore on these islands, not surprisingly, had large ecological ripple effects. Plants that were heavily grazed by the tortoises were able to grow unchecked. Plus, tortoises serve as seed dispersers, eating fruits in one place and pooping them out in another. With the tortoises gone, many plants had trouble spreading their seeds. This lack of dispersal was particularly problematic for habitat restoration efforts, because once native plant populations were reestablished, they had no way to get their seeds to new areas. Projects in which giant tortoises from Aldabra or somewhat smaller tortoises from Madagascar have been introduced to tortoise-less islands have been very successful in restoring functioning ecosystems that had been disrupted by loss of the native chelonian. Certainly, it would seem to be better to have some sort of giant tortoise, even if not the same species that was there before humans arrived, than no tortoise—and thus no dominant herbivore—at all.

———

This is conservation in the Anthropocene. Sure, we'd like to return the world to the way it was before humans took charge of running the planet. But that ship has sailed. We've got to do the best we can with the possibilities at hand, not to mention the resources at our disposal. And we have to recognize that the world is changing—as it always has—whether we like it or not.

Let's return to some of the examples I've discussed. In an ideal world, we'd return the thylacine to its place as figurative top dog in Australia. But barring a miraculous rediscovery of a living population or a nearly-as-miraculous resurrection by genetic cloning, that's not an option we have. The data at hand suggest that introduced foxes and cats are a bigger threat to Australia's surviving native fauna than are dingoes, and that dingoes might alleviate the damage they cause. Careful study is needed, but exploring the possibility that dingoes are a net positive for species survival and ecosystem functioning is warranted. Other cases where introduced species are filling the role of similar extinct species—wild horses, perhaps?—deserve similar consideration.

Does that mean we should intentionally introduce non-native species to fill the place of species we've eliminated? Biological control efforts have a very checkered history in which the impact of introduced predators has often had unexpected and highly detrimental outcomes, so extreme caution is warranted. But as the tortoise example illustrates, such efforts are worth considering.

What about armadillos? I'm not aware of any scientific study, but standard wisdom is that they have an enormous negative effect on local ecosystems by tearing up the ground cover and eating anything they encounter. They're spreading north on their own, but very likely this range expansion is prompted by warming climates and the reduction in large predators in North America, both the result of human actions.

My feeling is that if species are expanding their ranges by their own movements (as opposed to transport by humans), then we'll have to let this go. The whole world is affected by global warming—there's no way we can prevent species from responding as they have for eons. Indeed, if we don't let species shift their ranges, some are likely to go extinct. Get used to armadillos and marvel at how well they are doing in human settings (give opossums deserved respect as well). Rick Shine in his essay makes a similar point about how human changes to the environment have benefited some species, like the large populations of garter snakes resulting from the introduction of earthworms. It's just not feasible or desirable to try to undo those knock-on consequences of human environmental actions.

Finally, for introduced species that clearly are having a detrimental effect on native species, we must continue our efforts to humanely remove them whenever possible. A corollary is that we should endeavor to minimize the number of new introductions, given that it can be hard to predict which species will have negative effects. We are the stewards of this planet, and it is our responsibility to prevent species from going extinct from our actions.

Keeping a love of the wild alive will require us to love a different kind of "wild," one that has less fidelity to historical assemblages and older notions of ecological "integrity" and that is more about healthy and sustainably functioning systems. This approach will leave plenty of room for biodiversity, but in a worldview that doesn't see human actions as inherently and inevitably polluting some sort of sacred natural order.

4

Bringing the Wild Things into Our Lives

THE PROBLEM WITH CATS

Hal Herzog

"Something of every cat still vibrates with its wildness."
—PETER CHRISTIE

I RING Beth's doorbell. I want to hear about her cats—a pair of three-year-old brothers named Finn and Chatsworth. She and her husband Dick adopted them as kittens from a shelter. Beth opens the door and smiles. We have known each other since grad school, and she is glad to see me. So is Finn. He purrs and rubs against my leg. I reach down and scratch behind his ears, and he looks me in the eyes and purrs some more. Then he follows us onto the porch and plops into the chair next to Beth. She tells me this is how he spends most of his day, following her around. He is with her in the evening when she cooks dinner, and in the morning when she reads the paper. At night he sleeps in her bed. Finn likes Dick okay, but he adores Beth.

"Where's Chatsworth?" I ask.

"Chat is in the basement. He hides when a stranger comes in or if he hears a loud noise like the garbage truck. He's not comfortable in the human world. He never has been," she says. Then she tells me Chat makes life miserable for the furry and feathered creatures in their neighborhood. He sometimes brings home chipmunks, birds, and once, to Beth's dismay, a baby rabbit. But Beth rarely finds dead animals in her yard. She thinks Chat eats most of his kill.

Beth is deeply troubled by her pet's desire to hunt. "I wish my cat would not do it. I wish he would leave the other animals alone. But he won't." She has tried to reduce the moral burden that comes with living with an animal that only eats meat. "I don't allow him outside at dawn or dusk when cats are more predatory. And I don't let him out at night when he is most likely to encounter a predator. Coyotes have been seen around lately."

"Why don't you just keep the cats inside," I ask. After all, the ASPCA, the Humane Society of the United States, and the Audubon Society say cats should never be given unfettered access to the outdoors. She shakes her head and tells me she tried to raise Finn and Chat as indoor cats, but it did not work. "They were desperate to go outside right from the start."

Chat and Finn look identical to me, but multiple paternity is common in cat litters, and the brothers may have had different dads. Whatever the cause—genes, environment, the vagaries of random chance—their personalities are polar opposites. Finn is calm and likes hanging around people, begging for attention. Sometimes he will go outside for an hour or so, but he never leaves the yard. He shows little interest in hunting. His brother, however, is a born killer whose territory includes the whole neighborhood. Nervous around strangers, Chat likes to hang out—draped like a leopard in the Serengeti—on the limb of a tree.

———

The problem with cats is that they retain more inherent wildness than any other domestic animal. Of the sixteen thousand species of birds and mammals on the planet, only a couple of dozen were successfully domesticated. Nearly all the progenitors of modern domestic animals possess a common set of characteristics. They tend to be diurnal, docile herbivores or omnivores, social animals with dominance hierarchies. The ancestors of the domestic cat, however, did not fit the mold.

The modern house cat (*Felis catus*) evolved from small to medium-sized wildcats which ranged widely over Europe, Africa, the Arabian Peninsula, and parts of Asia. Based on genetic evidence, the current consensus is that my cat Tilly's forebearers were members of a subspecies of felids that inhabited North Africa and the Near East—*Felis silvestris lybica*. Unlike other successfully domesticated mammals, wildcats are nocturnal, territorial, antisocial loners. And, like every member of the cat family, they are obligate carnivores, creatures whose diet consists entirely of flesh. Wildcats began hanging around humans ten or eleven thousand years ago when *Homo sapiens* in the Fertile Crescent shifted from nomadic hunting-and-gathering lifestyles to the cultivation of grains. With the invention of agriculture came permanent settlements and the storage of food—wheat, rye, and barley—which attracted rats and mice, which in turn, drew wildcats. Their decision to hook up with humans paid off. While most species of felids are now threatened or endangered, the domestic cat has flourished. Estimates of the number of cats on Earth vary from five hundred million to a billion, and over 95 percent of them call their own shots. They choose their own mates and do not rely on humans for food.

The roles of cats in human lives have varied widely over time and across cultures. In ancient Egypt four thousand years ago, cats were worshiped. But during the Middle Ages in Europe, the Catholic Church linked cats with witchcraft. They were vilified, tortured, and killed en masse. In recent years, cats have come to rival dogs as the world's most popular pet. Modern human-cat relationships raise a host of psychological and ethical questions. Are some people drawn to cats like the predatory Chat because they represent our connection to the natural world? Does a gregarious cat like Finn who stays near Beth most of the day represent a different pull—our need for a furry friend? Is our desire to live with cats responsible for the decimation of wild bird populations and the extinction of entire species? Is forcing cats to spend their lives indoors psychologically harmful and bad for their health? Are there limits to how we should jigger the feline genome to fit human aesthetic preferences? Is it even ethical to own a cat?

Unlike dogs, modern cats were primarily shaped by natural selection rather than human intervention. But as witnessed by the growth of organizations like the Cat Fanciers Association, there is growing interest in the selective breeding of pedigree cats. Unfortunately, intensive selection for animals that conform to the arbitrary breed standards of competitive cat shows has resulted in creatures like the "peke-faced" line of Persian cats modeled after Pekinese dogs. With their big heads, large eyes, and flat faces, these cats are the feline equivalent of brachycephalic dogs. As a result, the peke-faced line of Persians suffers from afflictions such as breathing difficulties and dental issues that plague dog breeds like French bulldogs and pugs. Flat-faced cat breeds are susceptible to a host of other veterinary problems, including kidney disease, deficient tear duct drainage, skin disorders, and heart and blood diseases.

While some cat lovers are drawn to brachycephalic kitties with their big eyes, soft features, and large craniums—features of infants that evoke the "cute response" in humans—others seek pets that harken back to the cat's wild ancestry. Take the Bengal cat. It was created by crossbreeding domestic cats with Asian leopard cats (*Prionailurus bengalensis*). Repeated breedings between hybrids and domestics produced a line of offspring that look like they stepped out of the jungle. The success of the Bengal led to a slew of other feline hybrids. Among these are the Savannah cat (a serval X domestic cat cross), the cracat (a caracal X domestic cat cross), and the chausie cat (a jungle cat X domestic cat cross). While technically not a cross-species hybrid, the toyger is a line of domestic cats designed to resemble tigers. These animals do not come cheap—a top-grade toyger kitten can cost as much as $20,000 from Styled in the Wild, and Select Exotics will sell you an F1 hybrid Savannah cat kitten for $23,000. In addition to being expensive, hybrids are controversial. Savannah cats are banned in Australia where researchers estimated they could pose a risk to 93 percent of the country's threatened species of mammals.

In recent years, the primary role of cats in developed countries has shifted from pest control to companionship, and the feline hunting instinct is now considered a problem rather than an asset. The transition from a working predator to a family member reflects a trend the pet products industry refers to as the humanization of pets. The shelves on my local supermarket contain tins of gourmet cat foods described as "human quality." Chewy.com offers a line of "cute and sassy" cat dresses, and a plethora of video games have been developed for the enjoyment of cats. (My cat Tilly is keen on the iPhone version of Rat Tail Bandit.)

The humanization of cats was exemplified by the responses to a scale I once developed on the ethics of feeding different types of foods to pets. Pet food stores commonly sell frozen mice as food for pet snakes. In one of my classes, 80 percent of the students who took the scale approved of feeding dead mice to pet boa constrictors. Yet only 40 percent of them were okay with feeding dead mice to pet cats. In a subsequent discussion, I asked why most of them opposed feeding mice to cats; mice are, after all, their natural prey. One of the students appeared upset, and suddenly blurted out, "If my cat ate mice, she wouldn't be like me!" My student was in denial. She did not want to believe there was a reason her pet was equipped with retractable claws, eyes that see in the dark, and teeth that stab and slice.

The problem is that cats can take a toll on wildlife. Over 1,500 research papers were published between 2010 and 2020 on cats as predators. The most influential was a 2013 article in the journal *Nature Communications* by wildlife ecologists Scott Loss, Tom Will, and Peter Marra. They calculated that cats in the United States kill between 1.3 and 4 billion birds and between 6.3 and 22.3 billion mammals every year and that globally, cats are a serious threat to biodiversity.

Not everyone agrees. Other researchers argue that the claims about the impact of cats on native species is overblown and have generated a widespread yet bogus moral panic. But while the statistics might be fuzzy, pet cats in the United States and Europe do kill many millions, perhaps billions, of birds, mammals, and reptiles each year. An obvious way to eliminate the carnage our pets inflict on the wild things is to confine them indoors. Indeed, that is exactly what the American Veterinary Medical Association says its members should recommend to their clients. Indoor cat proponents argue that cats kept inside 24/7 typically live longer than outdoor cats and they are not susceptible to getting lost, poisoned, hit by cars, contracting contagious diseases, or being killed by predators. There are, however, negative health and psychological consequences to confining cats in the big cages we call our homes. Indoor-only cats are prone to urinary tract disorders, hyperthyroidism, dental disease, obesity, and diabetes. And they are more likely to destroy furniture, spray urine around the house, attack humans and other cats, experience separation anxiety, and exhibit repetitive behavior disorders.

Until recently, all cats spent at least part of their day roaming free. Full-time indoor life for cats was made possible in the United States by shifts in public attitudes, beginning in the 1980s, for example, the widespread acceptance of spaying and neutering, which reduces scent marking and sex-related howling. It was also facilitated by the invention of kitty litter and the development of commercial pet foods that better meet the special nutritional needs of cats. Today, between 70 percent and 80 percent of American feline companion animals are indoor-only cats. While the indoor-only ideal has gained traction in the United States, it has not been similarly embraced in other countries. According to owner surveys, 90 percent of pet cats in New Zealand spend at least part of their day roaming at will, as do 72 percent of cats in Denmark, 80 percent of cats in the United Kingdom, and 66 percent of Australian pet cats.

Animal protectionists are divided over the ethics of permanently confining cats indoors. PETA's stance is clear—every cat should be an indoor cat (though they admit lucky ones could have access to a fenced-in outdoor area). Ethicists are split on the issue. The moral philosopher David DeGrazia argues that the fundamental needs of cats for exercise and a stimulating environment preclude them from being confined to indoor life. He says that if you cannot allow your cat to roam outside— or at least be taken for walks on a leash—you probably should not have a cat at all. The philosopher and animal activist C. E. Abbate goes further. She argues that forcing cats to spend their lives indoors inevitably impairs their ability to flourish. Abbate believes that keeping cats indoors constitutes a moral failing on the part of their owners. Is forcing your pet to spend its day on a couch watching the birds flit about through a window a violation of its very nature? The ethicist Bernard Rollin thought so. Based on Aristotle, Rollin argued that every species has a *telos*—a unique essence we should not violate. Rollin wrote, "Fish gotta swim, birds gotta fly." In an email, once I asked Rollin if his position implied that "Cats gotta kill." His response was brief—"Yes."

Rollin's "gotta kill" principle, however, does not apply equally to all cats. James Serpell and his colleagues developed the *Feline Behavioral Assessment and Research Questionnaire* to study individual differences in cat behaviors. The Fe-BARQ is a behavioral inventory cat owners can complete online. Your cat's behavioral profile is then compared with thousands of other cats in the database. When my friend Beth completed the inventory for her two cats, Finn came out near the bottom in his interest in hunting. Chat, however, retained more of his *Felis silvestris* ancestry. Chat's Fe-BARQ hunting score was off the chart.

———

The humanization of pets has created a conundrum for modern cat owners. The more we think of creatures like cats and dogs as autonomous sentient beings with emotions and a sense of their own lives, the less right we have to desex them, confine them in our homes, and control every aspect of their lives. In short, the more we humanize pets, the less right we have to keep them as pets. And with cats, the ethical issue is compounded by their impact on wildlife. For me, the issue is personal. Tilly is a serial killer. Like Chat, Tilly maxed out on the interest-in-hunting item on the Fe-BARQ. Usually, her victim is a vole—sometimes partially eaten, sometimes an intact corpse deposited on the front porch. But there is the occasional chipmunk or towhee, and once, a juvenile garter snake.

Personally conflicted over living with a predator for whom indoor confinement is not an option, I turned for advice to Jessica Pierce, a bioethicist and author of *Run, Spot, Run: The Ethics of Keeping Pets*. I asked her if it is ethical for me to let Tilly roam outdoors. Her response did not let me off the hook—"I don't think there is any ethically clean, comfortable way to have a pet cat."

I agree. But there are a couple of strategies that might alleviate some of the moral burdens of life with a cat. John Bradshaw, the author of *Cat Sense: How the New Feline Science Can Make You a Better Friend to Your Pet* has suggested that lines of cats could be bred for low or no predatory instincts. In theory, he is right. In one study, researchers attached tiny video cameras to fifty-five pet cats whose owners allowed them outdoors. Over the two thousand hours of recordings, only 43 percent of the cats were observed to stalk prey, and they were generally not good at hunting. Just sixteen of the cats captured anything—and most of these cats only caught one or two animals. Another study found that about half of the differences between cats in their behavioral traits are attributable to genes. Hence, it should be easy to create nonhunting breeds of cats via selective breeding. But selective breeding for low predatory interest could have unforeseen consequences. For example, nonhunting lines of cats might eschew hunting because they are more likely to have visual problems, not because they have lost their predatory instincts.

The good news is that breeds of cat that show little interest in killing mice and birds already exist. The University of California at Davis researchers Benjamin and Lynette Hart asked eighty feline veterinarians to rate cat breeds on a variety of behavioral traits, including predation on songbirds. Persian cats and Sphynx cats had almost no interest in killing birds—they averaged a one on the ten-point predation items. Ragdolls were nearly as low. In contrast, domestic shorthairs like Tilly scored a ten—the highest of any breed. Serpell obtained similar results in his Fe-BARQ studies. These findings suggest people who want to keep their pets indoors should consider a breed like a Persian or Ragdoll rather than, say, a Bengal. (But Serpell also found that, unfortunately, Persians are not inclined to use litter boxes.)

Indoor-only cat owners can make life more interesting for their pets by giving them limited access to the outdoors. Cat researcher Mikel Delgado, for example, takes the moral burden of keeping her pets in captivity seriously. She spent $4,000 on a custom-made one-hundred-square-foot outdoor enclosed "catio" for her pets. And, because her cats like to climb, Mikel's living room includes a six-foot-high artificial cat tree with perches and cubby holes. While most indoor cat owners won't go to those lengths to ensure the happiness of their pets, inexpensive outdoor cat enclosures are commercially available. And some cats can be trained to go for walks on a leash.

Owners of outdoor cats can also take steps to reduce their pets' inclinations to kill. In a series of experiments, researchers at Exeter University reported that playing with cats an additional five to ten minutes a day reduced the number of prey items their pets brought home by 24 percent. And they found that increasing the amount of meat in cats' diets was even more effective at reducing hunting. Cats put on a high-meat wet-food diet showed a 36 percent decline in predation rates. Unfortunately, putting bells on cats did not affect the number of prey items brought home, and cats who were made to work for their dinner with puzzle feeders increased their kill rates by 30 percent.

Some of the ethical quandaries caused by our relationships with *Felis catus,* however, cannot be solved by meat-enriched cat food or keeping pet cats indoors. For example, 14 percent of extinctions of bird, mammal, and reptile species on islands have been attributed to predation by feral cats. The environmental ethicists William Lynn and Francisco Santiago-Ávila argue that ecological problems associated with feral cats are what policy planners call "wicked problems"—they have important consequences yet are nearly impossible to solve. Wicked problems involve multiple stakeholders with conflicting values and interests, and they do not have tried-and-true solutions. Indeed, the presumed solutions can have unforeseen consequences. Take the removal of feral cats from islands to protect native species. While these efforts are usually successful, they sometimes backfire. After cats were eradicated from an island off the coast of New Zealand, the breeding success of the Cook's petrel—a native bird—crashed because of an unanticipated increase in another predator, the Pacific rat. (When the rats were subsequently exterminated, petrel breeding success on some parts of the island tripled.)

———

Over the years, I have thought a lot about our morally complicated relationships with other species. I have made my peace with eating meat and catch-and-release fishing. Yet I remain more conflicted over living with a cat than any other animal issue. Indeed, the ten-thousand-year-

FIGURE 9. The heart of the wild—Tilly tries to bring down a giraffe on *Planet Earth II*. Photo by the author.

old human-cat relationship exemplifies the uncomfortable balancing act that comes with the human attraction to wild things and our penchant for trying to bring them under our control—this is a challenge running through more than a few essays in this book. Tilly was about three months old when she first showed up on our doorstep—well past the critical period when kittens easily become socialized to humans. As a result, she is generally indifferent to people, including my wife. She seems to only like me and Diane, our cat-sitter. Once, when I saw Tilly take down a hummingbird on the fly, I was torn between admiration for her sheer athleticism and horror.

But sometimes in the evening, she will jump in my lap, purr once or twice, and we watch TV together. She is particularly transfixed by nature shows like the BBC's *Planet Earth*—a little black panther sitting with me in a rocking chair, with her heart in the wild.

5

How Did We Get Here?

Peter H. Raven

ALDO LEOPOLD, an extraordinary visionary, expressed our place in the biosphere beautifully in his *A Sand County Almanac* (1949):

> It is a century now since Darwin gave us the first glimpse of the origin of species. We know now what was unknown to all the preceding caravan of generations: that men are only fellow-voyagers with other creatures in the odyssey of evolution. This new knowledge should have given us, by this time, a sense of kinship with fellow-creatures; a wish to live and let live; a sense of wonder over the magnitude and duration of the biotic enterprise. Above all we should, in the century since Darwin, have come to know that man, while now captain of the adventuring ship, is hardly the sole object of its quest, and that his prior assumptions to this effect arose from the simple necessity of whistling in the dark.

As we humans consume an ever-increasing proportion of what the global ecosystem is able to produce, we need to remember that we are only one of millions of species that comprise that biosphere and enable it to function. Despite this, we often speak as if we were somehow able to carry out our activities apart from the rest of life. Nothing could be further from the truth. We exist within ecosystems and are able to do so only because of the conditions that they maintain. Our lives depend completely on their sustainable functioning.

The continued functioning of ecosystems, in turn, depends on the species within them. We are, for the most part, unable to predict when so many of these species will have been lost to disrupt some or all aspects of the functioning of a particular ecosystem—but we certainly know that the limits are there! The problem is that when we view ecosystems exclusively from a narrow economic point of view, we have been used to assuming that the biological diversity that sustains them will increase with our demand for it, although this is clearly not the case. Biological replenishment has real limits, and we have already exceeded them generally. In other words, we face challenges that must be met by developing new kinds of thinking and taking actions of the sorts developed so effectively by Cambridge University economist Partha Dasgupta and his colleagues—ones that take natural productivity into account— or face inevitable collective disaster.

The stakes are high. In the early twenty-first century, a rapidly growing human population nearing eight billion in number is driving the first stages of a major extinction event that promises to be equivalent in scope to the one that took place sixty-six million years ago and changed the character of life on Earth. We have already substantially exceeded the carrying capacity of the earth and begun to overwhelm the productive capacity of the ecosystems on which we depend for survival. Thus, we are consuming an estimated 175 percent of the planet's sustainable productivity, cultivating or grazing about 40 percent of the earth's land surface, and affecting in one way or another every square centimeter of our planet. Already we, with our domestic animals, account for approximately 95 percent of the total weight of all terrestrial vertebrates! Our activities have already warmed the earth's temperature an estimated 1.1°C above its background preindustrial level. These activities are continuing, putting us on course to reach 1.5°C above the original base temperature—considered to be the point of no return in terms of controlling global temperatures—by the close of the current decade. If nations continue to fail to respond meaningfully to this problem, we may push the temperature up 2.3 to 2.7°C above the background level within a few decades. Such a level would wreck a large proportion of the world's agriculture, cause the widespread displacement of people, and drive more than half of the world's species to extinction. At least a fifth of these species face extinction now, with twice that many quite possibly in danger by the end of this century.

———

How did this come to be? Perhaps it's because we have failed to internalize Leopold's powerful, but still elusive, biological-ethical insight. The human evolutionary line split from the other African apes 6 to 8 million years ago and subsequently gave rise to the array of species that we call hominids. Among them, the first recognizable members of our genus, *Homo*, appeared about 3 million years ago. The first hominids to migrate out of Africa were members of the species *Homo erectus*. Invading Eurasia via the Horn of Africa about 1.8 million years ago, they gradually spread throughout most of Europe and Asia but not beyond. Neanderthals and Denisovans, exclusively North American in distribution, were presumably derived from *Homo erectus* in Eurasia; they persisted there, if only locally, until as recently as 33,000 years ago. *Homo sapiens*, our name for the species that comprises modern humans, originated in Africa some 300,000 years ago, reaching Europe by 200,000 years ago. They had migrated to Australia by 70,000 years ago and the Americas by at least 15,000 years ago, spreading rapidly to the southern tip of South America.

The *human* story, occupying a mere blink in time, is one that tells of our conquest of the globe. This conquest is leading inexorably to an extinction event with a magnitude comparable to that which occurred sixty-six million years ago, when the character of life on Earth changed permanently. Over the past million years and even earlier, humans were starting to exert an effect on the lands where they lived, especially through hunting and by burning to clear those lands for grazing. Our earlier ancestors, however, had relatively minor impacts on their habitats, essentially sustainable ones comparable to those of the other species with which they lived.

The key transitional point in human history occurred about eleven thousand years ago, when our ancestors began to cultivate crops and domesticate animals for food. In this way, they secured the year-round, dependable supply of food that had previously lacked. At the time the earliest agriculture was being practiced, only about one million humans existed in the entire world, and only about one hundred thousand, remarkably no more than the capacity of a single modern sports stadium, in the entire continent of Europe.

With their newfound ability to provide food to tide them through unfavorable seasons, however, people were able to live together in centers that grew over time into villages, towns, and cities, each with its internal social structure. The diverse professions that have come to make up our civilization arose when individuals were able to specialize for the first time in particular pursuits within an integrated complex society. Although people certainly created art and made tools earlier, among other activities, it took living together for prolonged periods for the many nuances of our contemporary civilization to be developed. Civic and religious leaders appeared in these societies, and they became increasingly stratified, as the distinctions between rich and poor were created and maintained. A key innovation was the development of written languages, which appeared about five thousand years ago in Babylonia (cuneiform), Egypt, and China. As these advances took place, the total human population increased to about three hundred million people at the time of the Roman Empire, some sixty-five million of whom lived within that Empire.

As cities grew ever larger, a smaller and smaller proportion of the total population was engaged in producing the food that we collectively needed. Over the past few centuries, the relative proportion of farmers in our total population fell still further regionally, as the practices employed became more intensive and the fields larger. In most of the world, however, single families continued to manage most of the farms. Most of the farms remained relatively small, continuing to include much natural vegetation and habitat for wildlife: fallow fields, coppices and hedgerows, wet meadows, and unmanaged ditches. Lightly grazed pasturelands can be rich in successional plant and insect diversity. Romanticism, in part a school of painting that dominated in Europe during the early 1800s, clearly demonstrates increasing preoccupation with the nature being destroyed. Even in the parts of the world that maintained relatively small fields, however, the ever-larger human population put increasing pressure on the remaining vegetation near them.

By three thousand years ago, large areas, for example, in Europe, had been altered by humans and began to present an appearance very different from what they displayed earlier. The process continued and was accelerated as worldwide migration established agriculture ever more widely in areas like the Americas, Africa, and Australia. The global population first reached a billion people in the first years of the nineteenth century, with our impact on nature becoming progressively more evident. As these trends continued, humans began to think more about the differences between natural areas and those that we had changed for our own purposes. At the same time that we loved our factories and smokestacks for what they represented, there was a growing appreciation of natural areas as something worth saving too. At the very end of the nineteenth century, Alexander von Humboldt's travels in Latin America, which he reported in magnificent detail, brought to many an exciting new view of communities and ecosystems, and emphasized the biological richness of the earth.

The billion people who lived in Napoleonic times doubled to 2 billion by the 1920s and climbed further to reach 7.8 billion in 2021. We estimate that another 2 billion people will be added over the next three decades, with the population nearly 10 billion by midcentury. As the contemporary population explosion continues, we shall face even greater tensions between feeding people and preserving a sustainable world in which the global ecosystem can continue to function with as many species as possible surviving. For many centuries, people have been aware of the disappearance of individual kinds of animals along with a few kinds of plants; they tried to find ways to preserve these. It was not until the late 1960s, scarcely years ago, however, that we began to recognize that our activities were in fact leading to the loss of large numbers of species instead of just a few prominent ones. As increasing amounts of the productivity of the tropics and of poorer nations generally were being transferred to a relatively few rich nations, it became obvious that we were losing huge numbers of species in the tropics without even knowing that they existed.

Judging from these trends, there will likely be very few tropical forests in existence by the end of the twenty-first century. Since the ratification of the Convention on Biological Diversity in 1992, more than a quarter of the tropical forests that were standing then have been cut, and logging seems to be continuing apace, by fits and starts, throughout tropical Asia, Africa, and Latin America. We know only a pathetically small amount about these ecosystems, having recognized only a small fraction of the organisms that compose them. Linnaeus, from his vantage point in northern Europe, knew fewer than twenty thousand species of organisms, but estimates of the total today run from ten million to twenty million, or even more. We have recognized and given names to no more than two million of these, but even for the great majority of them we know very few facts. No wonder that it is so difficult to make solid predictions about ecosystem functioning! We are only a single species among the millions that constitute these ecosystems, but we have assumed dominance over all the others, and we are continuing to forge ahead without regard for the consequences.

———

How, then, are we to integrate what is left of nature into a sustainable world that will continue to support the global ecosystem dependably into the future? As our numbers grew over the past five centuries from five hundred million to eight billion, we began to value what we were losing, to put aside parks and reserves, and to find ways to relate more closely to nature. We are still a very long way from understanding what we would need to do to build a sustainable planet for the future, but certainly most of us view trying to save enough of nature to sustain the world for centuries to come as essential.

Nearly a century ago, led by a young Soviet scientist named G. F. Gause, ecologists began to experiment with mixtures of species of microscopic organisms in artificial culture in order to try to understand stability in mixed ecosystems. What they discovered was that, if the mix were left alone, one species would eventually expand into dominance, overshoot the mark, and then decline rapidly. This phenomenon helps us to understand what we humans have set in motion in overwhelming the entire global biosphere, a development made possible by our complex social systems and flexible ways of adapting to most different terrestrial habitats. With collapse increasingly evident in many parts of the world, we are in the process of finding out whether our innate intelligence can save us from collective disaster. Or perhaps we shall discover what sorts of organisms follow us into world dominance!

Yet the point remains: we have gotten to this moment by acting as if there were no limits to what nature could provide for us, but along the way have learned that, in fact, there are clear limits that we have already exceeded. As we evolved, we formed ever-larger groups, with the lifestyle of most people changing from small bands of hunter-gatherers to city dwellers in urban areas, linked together as nations. By midcentury, more than two-thirds of us will be living in cities, all of their people obviously bringing most of what they need to support their lives from less crowded areas around them or far away. Few, if any, people who have studied our situation have concluded that as many people as live here now can exist here sustainably over the coming decades and centuries. This relationship means that we will continue to confront a worsening problem as we move through the coming decades: how to adjust to what we have, and not to use it up.

Stewart Brand formulated one of the clearest ways of delineating these problems half a century ago, in his *Whole Earth Catalog*. He wrote, "We are as gods and might as well get good at it. So far, remotely done power and glory—as via government, big business, formal education, church—has succeeded to the point where gross defects obscure actual gains." Our reaction, however, seems anything but promising. We do have brains, but our selfishness and persistent desire to enrich those whom we accept as "our people" over all others seem to drive us ever closer to disaster. In the face of these factors, it seems clear that we shall have either to change our ways or face massive catastrophe.

Certainly, moderating our consumption, empowering women and children, and finding ways to lower the number of people would be requirements of a more sustainable future, but the seemingly endless struggle between nations, regions, religions, and individuals to hoard more wealth than anyone else can get certainly is not working to build overall sustainability.

Some of us, especially those who live in industrialized countries, are worried about the destruction that they see around them. For most, however, there is simply an implicit assumption that what they need will come from elsewhere without limit, and that there is really nothing to worry about. Most countries measure their success by the wealth they accumulate, without regard to the effects of that accumulation on any other country or the world as a whole. In the US, for example, we are already using several times the productivity that our own lands are capable of supplying; there is obviously no way for all countries to do the same. We Americans look back with longing on the years immediately following World War II, when, owing to widespread global disruption, the United States controlled for a few years more than half of the world's economy. Every rich country, region, city, or individual in some way faces the same dilemma. Taught to compete, we continue to accumulate as much as we possibly can without regard to the needs of others. Indeed, rich nations like the US continue to drain off as much of the world's productivity as they can manage to appropriate for themselves, with some 10 percent of us living in countries that consume an estimated 40 percent of the planet's total productivity. This leaves little ecological room, and ultimately none, for the poor of the earth to improve their lot, and creates a kind of persistent disequilibrium that will ultimately bring us all down.

Historically, the more resources any group controlled, the better off they were. Ultimately, that is why roughly half a dozen individuals have come to possess roughly as much wealth as the poorer half of the total global population—some four billion of them. At the same time, a very high percentage of the world's people live in poverty, able to exist only by consuming whatever resources they can without regard to global sustainability. Overall, it will be necessary to reach a much higher degree of equality and social justice to attain lasting stability, and yet there does not seem any particular movement in that direction. In 1945, the founders of the United Nations certainly intended its formation to represent a step toward global sustainability through international collaboration, but both the UN and similar international organizations have tended to languish subsequently. Apparently "Me first," the most appropriate attitude when there are many fewer people than now, is deeply ingrained in our collective psyches, and we keep trying to find ways to outcompete "others," however the "others" might be defined. Sadly, this has tended to mean that when the collaborative organizations propose measures for the common good, the greed of individual nations tends to rip them apart. At the end of the day, each nation seems to want all that it can get, at the expense of other nations from which its wealth ultimately needs to be taken.

We have come face-to-face with the need to change and yet seem to be unable to act meaningfully in that regard. Looking at the problem intellectually seems to be insufficient to bring about the changes we need, judged from what we have been able to accomplish so far. In part, there seems to be a gap between the mode of reasoning of scientists and nonscientists. Scientists present hypotheses, which are tested and challenged until they are eventually regarded as well supported, modified, or rejected. Most people cannot follow the kind of reasoning involved, and in any case would probably not take the time to try.

At the end of the day, science does not tell us what to do: it outlines the consequences of what we might do. In terms of taking action, we generally accept a particular view based on what someone whom we trust has stated. Clearly, just as in advertising, hearing statements repeatedly, rather than judging them carefully, is what most often convinces people to accept them. What we must do to reach sustainability is to implement logic in planning the way we consume resources so that we can achieve the most positive results for the long run. In a fully cooperative society, we could theoretically accomplish this, but scientific thinking would need to inform the actions we decide to take. Can we learn to be noble, and to understand that acting in the general interest is actually the way to serve our own interests too? The time available is short, and the answer to this all-important question is by no means obvious.

Obviously, some individuals have had a major effect on the course of history; St. Francis, Gandhi, or Nelson Mandela for example. They have done so, however, based on a love for people and for the nature that supports us all. Religious and moral concepts have often proved important in guiding principled collective actions, although they too can lead to selfish conclusions and outcomes. In my own life, I have grown in my thinking from a deep interest in natural history to a concern about the disappearance of species on a catastrophic scale to a concern with regional and global sustainability. Certainly, we may find motives to move to sustainability even if we simply take the time to contemplate the beauty of nature. In this, we must move beyond nature's usefulness and the fact that we depend on it collectively for our existence to consider its beauty and the way in which it refreshes our lives with its incredible, seemingly endless beauty.

In this connection, the more we bring nature into our cities and the areas around them, the more likely we may be to understand, to prize, and to support it. Natural beauty is not simply something we enjoy, but a reflection of what is left of the ecosystems into which we evolved and which continue to support us now. All people, and all life, exist on a single planet, so that the need for managing the complex ecosystems that support us should be obvious.

E. O. Wilson's concept of saving "Half Earth" suggests that we organize large areas in a relatively natural state around our cities and beyond them; doing so would preserve the beauty as well as the function of these areas while at the same time conserving as much biodiversity as we probably can. Reasoning of this sort seems well suited to keep a love of nature and the world alive and, emphasized properly, to provide a reason for doing so.

Religions can be as hostile as nations in fighting and killing their enemies, but they have a deeper meaning that could lead their members to transcend their base instincts and focus on love for everyone. My belief is that we need to look for inspiration along these lines while striving to find rationality too. Love is the final answer, and winning our survival is the simple reason that we need love.

As botanist Hugh Iltis put it:

"If we love our children, we must love our earth with tender care and pass it on, diverse and beautiful, so that people, on a warm spring day 10,000 years hence, can feel peace in a sea of green, can watch a bee visit a flower, can hear a sandpiper call in the sky, and can find joy in being alive."

PART II

Wilderness, Wildness, Wild: Legacies and Liabilities

6

Why Does Anything Need to be Called Wild?

Kyle Whyte

I HAVE had the privilege to see some of Bernard Perley's brilliant cartoons in his presentations. Perley has published several series, including *Having Reservations* and *Going Native*. The series address critical issues endured by Indigenous people in life and in academia. There's a cartoon of Perley's I saw once that reveals for me one of the core problems that Indigenous peoples suffer when others bring up concepts of wilderness, wildness, and the wild.

The cartoon has three panels lined up vertically, all of which feature two Indigenous North Americans standing on an eastern seaboard coast circa the sixteenth or seventeenth centuries. They are watching the horizon as an unidentified object emerges from the distant ocean and comes closer to them. In the first panel, it's hard for them to tell what the object in the distance is because it's so small, and the Native persons exclaim "???". In the second panel, the object becomes bigger but is still indecipherable to them, and they exclaim "!!!". In the final panel on the bottom, the object is clearly a European ship, and one of the Native persons says to the other, "It looks like the wilderness just arrived."

For me, Perley's cartoon expresses the problem that Indigenous peoples never consented to some peoples' use of wild concepts as descriptors of Indigenous territories. I'll just use the term "wild concepts" here as an abbreviation for various concepts of wilderness, wildness, and the wild. The concepts are similar, though they have been used differently, depending on the context and the historical period. Typically, terms like "the wild" signal a culturally specific set of assumptions about how humans relate to land, plants, and animals. One of the assumptions is that such relationships can be understood in terms of degrees of human influence on ecosystems and on plant and animal behavior. On this understanding, wild lands or wild animals have been influenced little by humans. But in fact—and as some others have discussed elsewhere—there are various degrees or gradations of wildness in terms of freedom from human influence.

Numerous Indigenous peoples of what's currently called the US never really used anything like wild concepts to describe their relationships with land, water, plants, animals, and ecosystems. They were more concerned with respecting and enacting specific relationships of interdependence within ecosystems and with nonhuman beings, flows, and entities. Relationships of interdependence included diverse Indigenous cultures and societies. For example, scholars like Heidi Kiiwetinepinesiik Stark, Leanne Simpson, and Basil Johnston have shared rich stories and interpretations of Anishinaabe *treaties* with fish and furbearers in the Great Lakes region.

Having a treaty relationship with an animal community involves knowledge of the ecological interdependence of humans and animals. Involvement can be in the form of human ceremonies, harvesting, and caretaking practices that, at different times of the year, protect critical habitats. It involves recognizing that at other times of the year there are few if any interactions with these animals, and humans should stay away. There are even highly instructive stories of animals working together to address the fact that humans were breaching treaty obligations.

Treaties with animals motivate humans to act reciprocally, responsibly, and respectfully toward animals. Such behaviors are energized annually through ceremonies, educational institutions, and mentorship traditions. Empirically, maintaining treaties involves constant learning about how to maintain relationships of interdependence that support clean, safe, and healthy environments for all. Today, Anishinaabe have continued to grow their own knowledge systems for the sake of conservation. Many tribal nations and intertribal organizations have also taken up the use of newer tools from diverse environmental and climate science fields. Doing so is a contemporary manifestation of the treaty relationships.

When certain Indigenous people find themselves to have abused their treaties with animals, they take very seriously the ramifications of having literally breached a covenant. If we invoke treaties as a model for our behavior and learning, there's no need for any animal to be deemed wild or not. We can respect animals' self-determination, freedom, and ecological roles as part of the treaty relationships. Again, treaties are just one small example of Anishinaabe relationship interdependence within ecosystems and with nonhuman beings, flows, and entities. The knowledge, practices, and lessons associated with treaty relationships serve as motivationally significant and revisable frameworks for understanding how to engage in environmental stewardship, protection, and conservation.

Indigenous peoples did not create English language wild concepts. Nor did Indigenous peoples choose that wild concepts be applied to describe Indigenous relationships of interdependence within ecosystems and with nonhuman beings, flows, and entities. In the case of the US, colonists from Europe and settlers invented wild concepts. Colonists and settlers came up with wild concepts to describe ecosystems and environments that they didn't know anything about. They didn't know any of the histories. They had no empirical ecological knowledge of North America.

Anya Zilberstein describes an episode in the 1600s when Europeans were concerned about a global cooling crisis, which led to a significant reduction in temperatures in England. For the British, who were actively trying to colonize North America at the time, botanists were looking for plants that thrived in Canada and that could be transplanted to England. As the English name indicates, they believed the rice growing there to be *wild* rice, sustaining itself spontaneously in a colder climate. Given they never consulted the Indigenous stewards of wild rice, the transplantation efforts failed.

Such is one episode of the history of colonists and settlers' applications of wild concepts to North American environments, environments characterized by diverse and complex interrelationships with Indigenous societies over generations. Today, many settler Americans and Canadians derive value for themselves when they understand nonhumans and ecosystems as wild. The values may be spiritual, recreational, and economic. The values may relate to mitigating climate change and biodiversity loss. Settlers are sometimes quite adamant that wild concepts need to be used to describe some lands in North America—even though all lands have Indigenous living heritage. We're back to Perley's punchline: "It looks like the wilderness just arrived."

Wild concepts include all the terms that use the term "wild," including "the wild," "wilderness," and "wildness." They also include concepts of "savage," "unspoiled," and "terra nullius." In the initial uses of these terms historically and still in certain uses today, such terms denigrate Indigenous peoples' land management practices, land tenure systems, science and knowledge systems, and political sovereignty and self-government. The meanings and imaginations behind terms like "wilderness" or "terra nullius" played roles in justifying wrongful land dispossession and colonial conquest by the powers and institutions of US settlement.

"Wilderness," for example, has meant, in different periods, the belief that lands currently under US authority were largely uninhabited before European arrival and US settlement. Or, in cases where inhabitation by Indigenous peoples is acknowledged, Indigenous peoples' cultures, economies, and political institutions were deemed to be savage and hence in need of being assimilated forcibly into dominant US settler culture. The results associated with uses of wild concepts are appalling.

The alleged savagery was part of an insidious justification that some people invoked to force many Indigenous persons into assimilative and abusive boarding schools. US settlers suppressed and outlawed Indigenous burning practices and other forms of local and regional land management. They banned land-based ceremonies and cultural activities. They removed many Indigenous peoples from their own lands to make way for national parks and public lands. They advertised Indigenous lands as wild to attract recreational renters and home buyers. They denied

Indigenous peoples' rights in wilderness areas. Indigenous peoples continue to face risks and harm in wilderness areas, such as the resistance settlers posed to Bois Forte Chippewa harvesters being able to secure community sustenance in the Boundary Waters Canoe Area Wilderness in Minnesota.

In the case of a settler nation like the US, all places are infused with Indigenous heritage, meaning, and value. As mentioned earlier, wild concepts are sometimes used to suggest that there are places with little or varying degrees of human influence. The deeming of any place as having little or no human influence, then, always demands further explanation. In some places, the lack of human influence simply refers to a lack of Indigenous stewardship during the most recent decades or centuries. It's just land that hasn't been directly managed by diverse Indigenous peoples in a while.

In other places, some of the markers of wildness are the direct result of Indigenous management practices. Burning is one of the common management practices. In some cases, settler Americans have sought to use alternative processes (to fire) to maintain what they perceive to be the wildness or naturalness. Michael Johnson has discussed John Muir's role in ending Indigenous burning practices in Yosemite National Park and advocating for landscape design as a way to maintain the appearance of naturalness.

In other cases, Indigenous peoples maintain cultural, economic, and political traditions and protocols for management—but do so discreetly. For generations in the Great Lakes region, some Indigenous persons had to exercise treaty hunting and fishing rights secretly for threat of being harassed or accused of poaching by settlers.

It's important to be clear, then, as a matter of fact, that the access to and protection of wild places—whether wilderness areas or national parks—that many Americans take for granted, is a certain type of virtual reality for settlers. While there is virtuality in the paradigms humans use, this one can be willfully ignorant and resilient against revision. It's really confusing how someone could look at a place that has generations of history of Indigenous stewardship and land management and actually believe it to be wild. It's even more hard to believe that some areas that saw colonial massacres of Indigenous peoples are deemed to be wild.

Were it not for land dispossession and genocidal actions over centuries, the wild lands wouldn't be the ways they are now. Their accessibility to people wouldn't be the same. People wouldn't be able to access wild lands by invoking the same privileges of access and restriction that they assume to be normal. The maintenance of wild places today runs the risk of continuing to be dependent on excluding and restricting Indigenous self-determination and self-governance.

Of course, most people I know who use wild concepts in contemporary parlance place great distance between their intended meanings and the aforementioned history and associations with wilderness, savagery, and terra nullius. They want to express a meaning that is important to them for ethical reasons. They believe wild and wildness should be important to all for the sake of learning to live respectfully with the earth. Their meanings do not seek to pick up the colonial baggage and move it into the future (see Eileen Crist's essay in this volume).

In these ethically evolved meanings, wild concepts speak much to the importance of respecting and revering the independence of land, plants, and animals and honoring the biological, ecological, and climatic processes that should unfold on their own terms without disruption by humans. As other essays in this collection demonstrate, wildness can be perceived in many places, including urban areas. To respect wildness, humans should certainly recognize the benefits of letting wild beings and wild communities express their wildness. The benefits include recreational and economic ones.

There are moral dimensions to taking action to protect wildness, especially in terms of humans learning to treat non–human beings and communities in nonexploitative ways. Spiritually, there are experiences of great reverence that humans can have when they take advantage of opportunities to experience wildness firsthand. Certainly, such beliefs about the wild are not associated with land dispossession and colonialism. They are beliefs about a wildness that all people can appreciate and should respect.

Yet, many contemporary uses of wild concepts remain problematic. In some contexts, for example, wildness *still* serves as a vehicle for discussing ecosystems or nonhumans in ways that exclude an accounting

of Indigenous peoples' consent, self-determination, history, and knowledge. In these expressions, wild concepts have the effect of removing
Indigenous peoples from the histories of ecosystems and nonhuman
beings, flows, and entities.

Consider how in some accounts of wildness there are references to
events like the extinction of the passenger pigeon with no further reference to Indigenous peoples at all. In such references, it's as if there was
wildness popping out everywhere in North America when colonists
arrived and settlers established their permanence. It's the virtual reality of
wildness that Perley's cartoon speaks to. "It looks like the wilderness just
arrived." It is virtual reality to imagine settlers romping freely across
American lands exterminating the passenger pigeon. How did settlers get
such *freedom* to romp and exterminate?

It's telling that accounts of the destruction of wild animals and plants
do not reference how the exterminators gained the access, liberty, and
authority to do so. It's as if some of the major issues in American history
never happened, including conflict during the fur trade and early colonial
periods, treaty rights violations, forced relocation and removal, religious
and cultural bans, violence and racial discrimination, and assimilative
boarding schools.

The wildness of the passenger pigeon or American chestnut, and
their wild abundance from the perspectives of colonists and settlers,
creates an image of a historical trajectory in which these birds and trees
lived outside of human influence. They were just there, like the wild rice,
to be exterminated, appropriated, or left alone by settlers.

Beyond the issue of human influence, there is also the issue of cultural significance. When plants or animals are referred to as wild, there
is a cleansing of any other cultural beliefs that exist about them by Indigenous peoples, or even the possibility that there would, of course,
have been cultural beliefs in the first place. Does it not shock anyone
else that many of us have little immediate awareness of the cultural histories of the passenger pigeon or American chestnut?

Another issue with wild concepts is the idea that in places like wilderness areas, parks, or other public lands, one can experience wildness.
Often, I have heard people discuss wildness in this way. I once had a

friendly debate with a scholar of environmental ethics. The colleague claimed that he can only have a certain spiritual experience if he is confronted by wild nature. He gave examples of several areas where he has experienced wildness.

I pointed out that those areas wouldn't have been accessible to any of us today if it weren't for violent land dispossession of Indigenous peoples who depended on those same places—for generations. Indigenous peoples depended on those places for their own sustenance and for the thriving of their cultures, religions, and educational institutions. In the same friendly debate, I pointed out that Indigenous peoples today still do not consent to the current governance and access regimes of lands that were stolen from them.

Yet to my colleague, it was clear that the consent status of land doesn't matter. The consent status of land has no effect on his sensory and perceptual capacities to have a spiritual experience. I asked him whether the consent status of the land would affect his sense of enjoyment and spiritual rapture. I further asked him whether it was significant to his argument that the wild appearance of the land was actually the product of human influence in the form of genocide and violent land dispossession. He said quite clearly that his argument was true regardless of the consent status of the land and the recent genocidal history. He claimed that such considerations do not interfere with his emotionally charged spiritual experience and enjoyment of those places as wild places. I don't think he ever grasped the underlying assumptions in his argument.

Indigenous traditions have different philosophical views about spirituality and enjoyment. Consent matters. The consent status of land affects the spiritual character of peoples' experiences. Honoring recent history matters—including traumatic history. What does it mean when consent and trauma are detached from spirituality, emotional life, and enjoyment? Had there been time, I wanted to ask the colleague if there were any places in the world where the consent status and genocidal history of a place would be meaningful, or if it was just in the US where such issues were insignificant.

It's important to remember that the very "isolation" of some areas of US land is a product of genocide. Peoples' capacity to enter US land today in the way that they can is the product of there being no reconciliation that has yet been achieved with Indigenous peoples. Simply put, people should be careful to reflect on what assumptions are behind their images of wildness. What appears to be wild to some people is just that—an appearance. This is not an innocent appearance. For countless persons, including many Indigenous persons, wild concepts refer to lands that have been deeply disrupted by settler interventions. They are lands now prone to the problems of mismanagement, including wildfires and threatened fisheries.

Urban conceptions of wildness, while greatly valuable, are also morally problematic. Often wildness is contrasted with some industrial state of affairs in a city that is taken to be normal. Yet cities too, of course, are places that were built based on great upheaval of Indigenous peoples, and practices of land dispossession continue against many other groups of people through processes of gentrification, economic inflation, and labor exploitation.

If someone's attempts to disrupt city life are confined to carving out some narrow conception of wildness that operates in an urban setting, then that person's attention is heavily focused on a rather binary conception of what it would mean to make cities just and sustainable. Megan Bang and her colleagues have done important environmental educational work in the city of Chicago. They show that plants are often categorized as native or invasive. Calling a plant "invasive" attributes its migration to the plant itself. When in reality, it's the growth of the US as a colonial, capitalist, and industrial nation that is invasive, and is the cause of the plant's migration. Similarly, calling a plant native or indigenous obscures the diverse roles that humans played in maintaining habitats for "native" species.

Of course, Indigenous peoples today are not strangers to using wild concepts. There is the Native American Fish and Wildlife Society and the Great Lakes Indian Fish and Wildlife Commission. The Salish-Kootenai tribe manages its own wilderness area. So too does the Blue Lake Rancheria and the Yakama tribe, though in some cases the term "wild" is not used. Numerous tribes and first nations have their own tribal parks now. Some of my favorite authors and many Indigenous land managers I know use the term "wild" on an everyday basis. Often, they use "wild" in contrast to the ways in which settler Americans confine animals and ecosystems through pet relationships, industrial agriculture, and laboratory studies.

In Indigenous uses of wildness, "wildness" refers to certain characteristics of specific relationships involving ecosystems, humans, and nonhuman beings, flows, and entities. It is not a denial of any human influence. Rather, Indigenous conceptions of wildness assume the inevitability of this influence. What matters is the *type* of influence, that is, the type of *relationship*.

Tribal parks, for example, offer recreational opportunities for visitors. But the histories that park staff and community members tell and the available activities engage in an understanding of something like wildness differently. Instead of wild as something beyond human influence, Indigenous peoples emphasize a deeper history of connection between humans and *wild* plants, animals, and ecosystems. Human societies learned over generations about environmental management, including many hard lessons from past mistakes.

Indigenous peoples' languages, cultures, and land stewardship practices are important pieces of the history of "the wild." In Indigenous peoples' histories, there are some places that people engaged with directly all the time, others less so, and then certain places where it was widely accepted that people should not bother.

For Indigenous peoples in contexts I'm familiar with, the consent status does matter in terms of whether particular places can be spiritually meaningful. In such a relationship, the self-determination of humans and animals is respected, creating relationships of reciprocity. In these understandings, there may be some "wild" plants or "wild animals" or "wild places" that humans interact with rarely. Yet there may be others, so-called "wild," that are being engaged by humans constantly.

A medicinal gatherer who refers to medicines as wild plants is neither suggesting there is no human influence nor that there are differing degrees of human influence. Instead, that gatherer has special knowledge about the habitat of the plants, how to monitor the habitat, how to harvest and use the plants sustainably, how to cooperate with other harvesters, and how to teach others about the importance of conservation for the sake of future generations. Protocols of who can visit certain places and when are bound up with management strategies that humans very much believe they are responsible for. Consent is critical especially when multiple gatherers and other stewards may harvest nearby or share the same places.

Importantly, consent is a powerful way to enact one's relationships within ecosystems and with nonhuman beings, flows, and entities. The belief that humans are responsible for respecting the consent of nonhumans motivates an empirically critical mode of reflection. It is a critical mode of reflection that moves us to constantly question whether our stewardship practices are the best ones in terms of morality and sustainability.

When I think of the ways in which Indigenous persons have used wild, it's just very different depending on the community. Moreover, those uses of wild concepts are often expressed in passing. No one hangs their hat on the significance of wildness or the wild as major philosophies. Rather, they are English language words that, on occasion, help to express particular meanings if the interlocuters understand the context. What is much more important is the specifics of the relationships in question. The specifics involve the knowledge humans have about how to do their part to protect the self-determination of ecosystems and nonhuman beings, flows, and entities.

Sometimes deep reciprocity is needed, a constant tending of landscapes. Other times, true respect or even trust requires that we leave things alone. In each instance, there are reasons why certain relationships are treated in certain ways. The reasons are very specific. In the communities and places I'm acquainted with, empirical grounding in experience is crucial for framing the reasons humans give for the decisions they make about conservation. Often reasons are encoded within Indigenous people's languages, ceremonies, social worlds, and political systems. In a lot of the contexts in which I work, Indigenous persons use careful language to describe nonhuman relatives, whether ecosystems or nonhuman beings, flows, and entities.

In my experience, many Indigenous persons are quite open to reflecting on all the different sources of knowledge. When seeking to express plant knowledge, we may use diverse terms, including Indigenous, common English, and scientific names. I know many Indigenous persons who regularly invoke all of these names, including invoking them all in the same conversation. Being able to speak widely about the different names generates a respect for the overall history a plant has been through, from its deep history to the more recent history with US settlement. Terms like "wild" or "wildness" just can't do that by themselves. And when they are invoked as major descriptions of any plant, flow, animal, entity, or ecosystem, they do more to erase and obscure than they do to enlighten.

In my view, we should recommend that "wildness" be discarded as a term or concept. The only time that wild concepts seem useful to me is when they serve as an occasional convenience in our conversations. Many of us will continue, in our daily speech and writing, to use "wildness," "the wild," "wildlife," and "wilderness."

This book is an example of why wild concepts are not going away. In fact, one of its driving questions asks how we can keep a love of the wild alive. So, I don't think anyone will stop using the wild concepts. "Wild" and its variants are everyday terms. They are terms that many persons, I believe, find convey their intended meanings depending on the context. Certainly, wild concepts can be offensive terms with horrific pasts, including their direct implication in genocide, regardless of their intended meaning. But in other contexts, they will likely retain an everyday sense that depends on the intended meanings and contexts.

Yet, at the same time, does it even make sense to divide up the world as wild and not-wild, or even in terms of degrees of wildness? I am unclear on what advantages accrue to me or the planet when I use the term "wild" to understand certain places or living beings, whether from a traditional, scientific, or ethical standpoint. Moreover, given the painful history and the current suffering associated with the protection of wild places, reconciliation with Indigenous peoples risks being greatly hindered by any further proliferation of concepts of wildness. I think the issue of reconciliation is critical here. If more people today are committing to reconciliation, that involves undoing erasure of Indigenous culture, history, self-determination, and governance. I'm just not familiar with any contexts where wild concepts would bubble up from attempts to undo the suppression of knowledge about Indigenous peoples.

My solution is that people should look to how Indigenous conservationists today, throughout their own tribal parks and wilderness areas, are focusing carefully on language and history, and finding ways to express relationships that can motivate spiritual experiences and stewardship knowledge without having to ignore the consent status of any lands or the colonial genocide that has occurred.

And this solution is not outward looking only. Indigenous persons in the US context must have this conversation internally. Increasingly, tribal nations are being wooed by states and other entities to conserve their lands for the sake of carbon credits, often buying back stolen land through the proceeds. Yet as tribal nations convert recently reacquired lands and transform them into places of tribal and traditional stewardship, it needs to be acknowledged what industrial activities generate the proceeds and what communities are negatively affected by the continuance of the industries supplying the revenues.

Indigenous peoples have watched their entire set of territories be colonized and then converted rapidly to cities and towns, highways, industrial agriculture, wilderness areas and parks, and energy and transportation infrastructure. As in Perley's comments, colonists and settlers brought the wilderness! All places in the US context have had heavy recent human influence, stacking onto generations of Indigenous human influence, whether that Indigenous influence was intensive in particular areas, or about the sacredness and ecological benefits of letting certain beings or places be. For me, the answers to the climate change crisis and the coupling of environmental and social justice cannot be found in paradigms that view the world as either human influenced or nonhuman influenced, even if such paradigms admit various gradations between. We must attend to the particular relationships within ecosystems and with nonhuman beings, flows, and entities if we are to foster justice-based collective action toward protecting ecosystems and averting a climate crisis.

7

Affirming the Wilderness Ideal

Eileen Crist

THE WORD *beach* seems innocent enough, delightful even, to those who love visiting the sea to walk, swim, and soak up some sun. Yet words are rarely as innocent as they may appear—they tend to come freighted with images, connotations, and affect. "Beach" maps well with an experience I had on a Greek island. Upon following stone steps sign-posted "To the Beach," a 90-degree turn revealed a sand area stocked with umbrellas and long chairs with people tightly packed side-by-side sunbathing, reading, or staring at their phones. The setup also allowed people to order drinks from waiters who lingered about tensely (being overdressed) in the background. There were so many people on the beach that newcomers had to wait for a spot to become vacant, and then pay ten euros for it (drink or no drink). I think of such experiences as *Anthropocene* experiences. The Anthropocene, in that sense, is less about a geological epoch than an experiential flavor of human takeover.

I am not a misanthrope. I do not begrudge people their beach experience of enjoying a drink, scrolling through social media, and occasionally gazing at the sea's blue bounty. Yet neither can I help thinking—being that ecology has my undivided attention—of who and what were there in prior times (historical or deep), before the beach became human-*occupied* as opposed to being graciously *shared* with nonhuman others.

Then my thinking gets around to the word "beach," which I notice is loaded: its images, connotations, and affect are colored by the human element. To be sure, one can come across an empty beach. Yet the beach always lies in wait for the sunny or weekend day when it will be inundated with people. In some places in the United States, beaches are also inundated by four-wheeling SUVs. You can catch such an Americana scene at some beaches of North Carolina's Outer Banks, which offer the "fun" option of driving a military-grade vehicle (one's own or rented) on the beach, packing the sand down and crisscrossing it with fat tire tracks. If it strikes those recreationists' fancy, they can park their SUV right at the water's edge, enjoy barbeque and beer, while on occasion feasting their eyes on the Atlantic blue-gray horizon. I might pointedly add that if such visitors have been coming to the same beach for a few years or decades, they may vaguely wonder where all the shorebirds went.

Affirming the wilderness ideal, in the face of the seized human ownership of the world, should not appear as an outlandish or radical standpoint. Affirming the wilderness ideal does not entail loathing people and their insane entitlements, nor wishing that they'd just all go away. Affirming wilderness simply says this: There's a wide wild world out there filled with all sorts of beings, most of them not human. Let them be free to be and to become who they are, to make worlds together, to rest *their* gaze upon the expansive beauties of this world. (If you do not believe that animals love to enjoy a view, you have not observed enough animals.)

There is a place by the ocean that my husband and I like to visit in Costa Rica. You could call it a "beach," but the word sits uneasily with the topography. I think of this place as a wild shore. It is a spacious expanse of sand, decorated with driftwood, tree trunks polished by the elements, and inlets and pools of ocean and river waters. This shoreline is *constantly* changing: it changes with the seasons (two in Costa Rica, green and dry), with the vicissitudes of weather, and, of course, with the tides. It never stays the same and sometimes it becomes completely inaccessible to people. It's like a wild canvas that some crazy-wisdom god keeps painting anew. Or, like impersonal feng shui ceaselessly reinventing and rearranging the place's features. There are some things you can mostly count on to be there—such as a jungle patch near the water with an extended family of scarlet macaws making a racket. Vultures are also omnipresent, taking in the horizon while waiting for the Pacific to bring them some "seafood" such as carcass of sea turtle or shark.

The expanse of this magical ecotone where seaside and jungle intimately and endlessly converse is framed by two rivers. River to river it is a laborious fifty-minute walk, with the Pacific on one side and tree-covered, cloud-smothered mountains on the other. The rivers are crocodile habitat and on the treetops, aside from birds, you might catch sight of a sloth. Sloths tend to perch on the limbs with a wide-angled view, but as far as I can tell what sloths mostly do is meditate.

Many of the world's shores (along with all its other places) should be wild and free. We can visit them, though there will be times they should be off-limits, when birds and sea turtles (in the case of seashores) are nesting. Why do we think that the world belongs to humans, to do with as we please, to make it and call it the Anthropocene? Wilderness lovers simply point out that it is better for most of the world to be a canvas painted by a crazy-wisdom god.

———

Wilderness came under fire, as both idea and reality, in the 1990s through a series of academic publications that became broadly influential. With doubt and aspersion cast upon wilderness, many began to regard it as an obsolete entity: suspect for its Western ideological origins and unsound for driving a wedge between humans and wild nature. The undermining of the wilderness ideal—as sizeable, relatively undisturbed natural areas in the world worth admiring, preserving, and restoring—has done a profound disservice to the natural world: it undercut environmentalism's credibility to advocate in strong defense of planet Earth.

The meaning of wilderness is neither a projection of the human mind nor a matter of cultural opinion. It is Earth's primordial manifestation of nature. Wilderness refers to autonomous, expansive, connected, and ever-in-flux natural areas, where a diversity of living beings are interdependent in relations of symbiosis, competition, predation, affection, tolerance, and mutualism. Wilderness creates the richest manifestations of life on Earth in terms of variety of life-forms, ecological complexity, rife evolutionary potential, hybrid living-nonliving phenomena (like soil and coral reefs), emergent and spatiotemporally extended phenomena (such as ecotones and migrations), and nonhuman behavioral repertoires, cultures, and minds. Terrestrial and marine wilderness is dynamic and regenerative, able to absorb and bounce back from formidable natural disturbances such as wildfires and hurricanes. What's more, in wilderness large-bodied wild animals can continue to exist and evolve: healthy populations of large carnivores and herbivores require bigness, habitat connectivity, remoteness from human presence, and freedom from unwarranted intrusion. What remains today of wilderness are natural expanses that are unfragmented, or minimally fragmented, by the human technosphere.

Critics have made much of the claim that "wilderness" is unjust in ostensibly excluding human beings. I believe the matter is not so simple. Wilderness may exclude *or* include humans, depending on the context in which we are contemplating human presence in relation to the wild. To be sure, human presence is not constitutive of the reality and meaning of wilderness since the latter is independent of the human variable: wild nature preexisted humans and will outlast the sojourn of our species. Yet wilderness also clearly includes humans, for it birthed the human species. Human beings can remain integral with wilderness as long as they sustain reciprocal relations with wild nature in ways that retain its dynamism and regenerative qualities. This has been, and often remains, the case with Indigenous people (with some exceptions). Therefore, wilderness does not exclude humans with cultural traditions (material and ideational) that foster belonging with, and participating in, the orchestra of life that surrounds them. However, wilderness emphatically (and essentially by definition) excludes humans with traditions bent upon the appropriation, subjugation, destruction, and aggressive management of the wild. Clearly, modern humans equipped with a mindset that defines the natural world as "resources," and with technologies invented

FIGURE 10. A photo we took when my husband and I visited this tree in 2021. It grows on the lovely property of Tilapias La Cascada, which is in the Puntarenas province of Costa Rica just south of the town of Ojochal. This tree (*Ceiba pentandra*, more informally known as kapok) is estimated to be over two hundred years old and is the third-tallest tree in Costa Rica. It has a magnificent presence. Photo by the author.

to exploit those so-called resources, are inimical to the wild. Their presence and activities destroy wilderness.

The academic offensive against wilderness as a "dubious human invention" occluded clear sight of wild nature's primeval manifestation. As a result, sight of nonhuman beings and processes that require wilderness to flourish and evolve—not to say remain in existence—was dimmed. The loss of sight of wild nature's original standing did nothing to discourage, and arguably buttressed, its continued obliteration. The understanding of wilderness as nature's aboriginal design (having preceded our existence and almost certain to outlive us) worked as a stronghold against the ascendancy of *relativism* with respect to wild nature's meaning. Relativism was exactly what wilderness critics propounded by foregrounding the anti-essentialist idea that the multiplicity of cultural perspectives on the natural world is exhaustive of nature's meanings. Once the doors to relativism swung open, not only did different cultural conceptions of (wild) nature become *à la mode*, but the entire topography of what nature *could* mean was assumed to be an anthropological affair. The ascendancy of relativism encouraged abandoning any baseline of wild nature's primal expression that is independent of human perspective and against which human impact might be gauged. This development attenuated cherishing and guarding the inherent integrity of the wild nonhuman world.

Even as wilderness critique vitiated the defense of wilderness by stripping it of independent signification, by the same token the critique effected something more subtle but equally damaging. It reinforced the received belief that all meaning, including the meaning of (wild) nature, stems from human mind and culture. Wild nature was denied inherent meaning, denied displaying its own standing, experience, and value. The evisceration of wild nature's intrinsic meaning helped deliver the fate of wilderness into human hands for any decision-making, manipulation, conversion, extraction, destruction, killing, management, or "improvement."

Alongside these downstream effects of wilderness critique—of weakening wild nature's defense and reinforcing the notion that only humans determine meaning—the anti-wilderness thesis paved the way for the emergence of "the new environmentalism." The new environmentalism braced the human-centered understanding of wilderness propagated by its critics: this legitimated further erosions of wild nature and validated wholly subsuming the natural world under various human schemes. For the new environmentalism, the most important mandate became the sustainable use and management of wild nature and the fair allocation of "natural resources" among all humans. The idea and reality of self-willed nature, brimming with intelligent agency and intrinsic value, went largely by the wayside.

———

A widely circulated 2011 paper titled "Conservation in the Anthropocene" heralded the new environmentalism. Authors Peter Kareiva, Robert Lalasz, and Michelle Marvier proclaimed that "the wilderness ideal presupposes that there are parts of the world untouched by humankind. The wilderness so beloved by conservationists—places 'untrammeled by man'—never existed, at least not in the last thousand years, and arguably even longer." New environmentalists thus echoed the requiem for wilderness, in conjunction with enunciating the arrival of epoch Anthropocene in which humans have (allegedly) become the decisive planetary force. "Nature no longer runs the Earth," according to new environmentalist Mark Lynas. "We do. It is our choice what happens here."

The assertion that humans have completely altered and assumed control of the planet is further underscored by vocabulary describing Earth as "domesticated," "used," and filled with "working landscapes," "anthromes," and "novel ecosystems." Emboldened by the first wave of wilderness debunking, new environmentalists urge that environmentalism now calls for a people-friendly identity; conservation efforts should be undertaken in the controlling context of human needs and demands. Echoing the relativism promulgated by wilderness critique, the reality of nature (wild or otherwise) became one to be decided among different *stakeholders*, a contemporary buzzword which refers exclusively to humans.

The turn against valorizing wilderness—attended today by a tacit proscription against even using the word wilderness if unbelted by scare quotes—propelled the idea of "ecosystem services" to the foreground as the human-friendly rationale for conserving relatively intact natural places. The argument for some level of wild nature protection has become pragmatic and economic: some natural areas are worth "more" to human well-being and profit-making when left standing than when converted or destroyed. Cost-benefit models are deployed to demonstrate, for instance, the services provided by forests, coral reefs, or marshlands. Former CEO of The Nature Conservancy, Mark Tercek, expressed this paradigm shift in an article titled "Money Talks—So Let's Give Nature a Voice." "Thinking about the value of nature," he writes, "leads to other ways of thinking familiar to business analysts. For example, concepts such as *maximize returns, invest in your assets, manage your risks, diversify*, and *promote innovation* are the common parlance of business and banking. These are rarely applied to nature, but they should be." Conceptualizing (wild) nature within a human-service and monetary paradigm reinforced the blow against the ideal of protecting nature *for its own sake*. Tossing that aspiration into history's dustbin—within two decades of the initial wilderness critiques—was likened to waking up from a fantasy. Environmentalism could finally join the so-called real world where tangible human benefit and money-in-the-bank are the bottom line.

Indeed, Peter Kareiva and Michelle Marvier (in a paper titled "What is Conservation Science?") counsel that, "realism is in order." The world will never again teem with wildlife, they proclaim, especially not with big animals like grizzlies, wolves, and sharks. We must resign ourselves to the reality that in the Anthropocene biodiversity will dwindle. We are reassured, however, that such losses are not fatal blows. For example, the same authors in the "Conservation in the Anthropocene" paper (cited above) write, "Passenger pigeons, once so abundant that its flocks darkened the sky, went extinct, along with countless other species from the Steller's cow to the dodo, without catastrophic or even measurable effects." The academic discrediting of wilderness launched in the late twentieth century tilled the soil for such early twenty-first-century nonchalant assertions. The new environmentalism took wilderness debunking to its logical conclusion: they urge humanity to let go of preserving (another censured word) and recovering a wild world rife with diverse and abundant life. They also counsel dropping the sentimentalism of grieving for the passenger pigeon, Steller's cow, dodo, and countless others driven prematurely to extinction. *Let's be real*: we cannot even measure the effects of their demise.

The encouraged resignation to the downward spiral of wilderness, and to the evanescence of the erstwhile biodiverse world it encompassed, extends even to the event of an anthropogenic mass extinction. The blasé mindset of relativism—wherein nothing of objective priceless value in the natural world can be lost since everything is a matter of human perspective—has perversely morphed into endeavoring to present a rosy side to an imminent human-driven mass extinction event. "Every other mass extinction led to a burst of profound evolution afterward," states Kareiva. Chris Thomas agrees: "The flip side of a new great extinction would eventually be a new evolutionary explosion. A new genesis, if you like." Vacating life-filled wilderness of inherent signification has arguably facilitated such bone-chilling apathy toward the fate of wild nature.

New environmentalists also second wilderness critics in countering the view that human presence mars and endangers wild nature: they regard "pristine wilderness" as an anti-human idea, for excluding humans from wild nature and, more generally, for casting humans as detrimental to the planet. New environmentalism insists we see ourselves as just another force of nature in the flux of life's history. We should embrace humanity's nature-molding gifts. Yes, sometimes humans are destructive, but we also have creative agency in engineering landscapes, shaping ecological niches, and moving species around into novel configurations of life. In alignment with reframing the human impact as the most recent episode in Earth's natural history, new environmentalists spurn catastrophizing about the ecological crisis. Forecasting collapse and displaying a doom-and-gloom attitude are passé. Instead, an upbeat outlook is recommended. The Anthropocene is not a dreaded or dangerous outcome but an age "ripe with human-driven opportunity." No more "woe to me and shame on you," for "a good Anthropocene is in our reach."

Naturalizing the domination of nature by making it qualitatively continuous with Earth's 3.8 billion-year natural history works to legitimate the obliteration and dislocation of countless species and the wholesale takeover of ecosystems and biomes. Additionally, suggesting that human planetary dominance can yield a stable and even flourishing "epoch" fosters a false sense of security for many people, who are not paying close attention to the gravity of our ecological predicament. Arguably, both these new arguments are riding the coattails of wilderness critique, which seeks to dissolve any boundaries between wild nature and human presence—including boundaries warranted by respect and reverence for nonhuman life.

New environmentalists welcome a human-landscaped world—with needed correctives to safeguard civilization—checkered by agricultural landscapes, human settlements, industrial plants, extractionist operations, and an infrastructural grid of highways, roads, satellite technologies, pipelines, cellular networks, hydropower dams, power lines, ship lanes, and underwater cables. Human beings, according to the new environmental paradigm, may well be able to prosper in a world dominated by the technosphere in which wild fish and wild forests, teeming wildlife and coral reefs, have vanished. New environmentalists embrace modern technological and lifestyle trends, and urge shedding the technophobic anti-modern image that has hounded environmentalism.

Endorsed technologies include nuclear power, genetic engineering, mega-dams, "precision agriculture," de-extinction, and (more tacitly) climate geoengineering. The pro-technology stance embraces technological gigantism and invasive interventions that nature lovers have long eyed with dread. Technology is sweepingly eulogized for its avowed benefits for the developing world, for its ostensible solutions to formidable ecological challenges, and for its trailblazing of humanity's onward historical march. "We need a worldview that sees technology as humane and sacred," implore Michael Shellenberger and Ted Nordhaus. Others in the new environmentalist platform express a consonant view of technology as empowering humanity to continually break natural limits and thrive. "Since prehistory," Erle Ellis notes, "human populations have used technologies and engineered ecosystems to sustain populations well beyond the capabilities of unaltered 'natural' ecosystems." "Who knows," he muses, "what will be possible with the technologies of the future?"

The technophile turn of the new environmentalism also has direct ties with the disparagement of wilderness as a primordial realm that deserves and demands human restraint. A strong conception of wilderness—as expansive natural areas free of human technosphere and excessive interference—obviates against the indiscriminate embrace of modern technology. This view does not reject technological development, which is indeed a gift of the human species, but calls for a far more nuanced and critical perspective on it, most especially when it comes to the violent infringement and penetration of infrastructures and other technologies into the natural world.

Wilderness critique worked as the new environmentalism's launching pad. The human centeredness tacit in the former became the platform explicitly championed in the latter. As noted above, new environmentalists enlist a services and monetary idiom to protect some natural areas for human benefit. They also incite humanity to resign itself to species extinctions, and even to a mass extinction event, since "life goes on" in their wake. New environmentalists are upbeat about a human-dominated world in the Anthropocene and embrace modern technology's march as human destiny; they offer no caveats, nor suggest any restrictions, to the technosphere's unending sprawl. They stereotype the views of pro-wilderness environmentalism as antiquated, romantic, misanthropic, or Luddite. As conservation biologist Michael Soulé protested, according to the new environmentalism loving wild nature for its own sake is supposed to be "a dysfunctional antihuman anachronism."

The tenets of the new environmentalism ratify the retreat from saving wild places and beings for who they intrinsically are and in recognition of their right to exist and thrive on planet Earth. New environmentalists seek to entrench what wilderness criticism had earlier foregrounded: anthropocentric environmental ethics, politics, and conservation. It is always humans among humans who construct, decide, and negotiate the values and uses of the natural world. *Nonhuman nature has no voice of its own.* Wilderness criticism, and new environmentalism in its wake, thus enshrine a founding principle of Western civilization: that humans are sole creators of meaning and that humans are sole stakeholders when it comes to the fate of the natural world. Upholding that age-old

Western anthropocentrism, which ideologically bankrolled nature's destruction over centuries and millennia, it is deeply ironic that wilderness critics repudiate wilderness as a putatively *Western* ideal.

———

As I see it, this anachronistic stereotype of pro-wilderness environmentalism is not just damaging, it is spurious. There is nothing backward looking about cherishing and defending nature for nature's sake alone. Indeed, that perspective reflects a revolutionary critique of the human-centered worldview (most especially developed by Western culture) that has long silenced nature's inherent voice and capitalized on nature's destruction. Classical environmentalism set forth the historically groundbreaking idea that wilderness has, to paraphrase political scientist John Rodman, its "own existence, character, potentialities, forms of excellence, integrity, and grandeur." Human beings, armed with a supercilious, self-ascribed sense of specialness and entitlement, have neither right nor warrant to subjugate that wild world.

The forward-looking vision of classical environmentalism has found a new level of clarity in our time (partly through tension with the new environmentalism) in the emergence of the Rewilding movement. This movement is culturally diverse encompassing projects spearheaded by activists, NGOs, conservation scientists, citizens, and writers around the world. The Rewilding movement advocates for the preservation of existing wilderness areas and for the restoration of degraded nature into wildlands, thus substantially expanding protected nature overall. The ultimate advocacy of the Rewilding movement is to create an ecological civilization in harmony with wild nature. This vision is in tension if not outright opposition with an Anthropocene (and new environmentalist) imaginary of a managerial technocratic civilization engaged in sustainable resource use, mega-technological enterprises, whole planet surveillance, and assorted "damage control" schemes on a planet dominated by and for people.

Rewilding is a type of ecological restoration with the goal of returning natural areas to wild self-governed states. Understanding wilderness as baseline and blueprint does not mean curating historically specific ecological constellations in "museum-like" states. It refers to reinstating conditions wherein the dynamism, regeneration, and creativity of wilderness can be expressed, eventually without human assistance or management. Size is important for the flourishing of wild nature. Larger natural areas can harbor and retain more species, subspecies, and populations of plants, animals, and other life-forms. Relatedly, bigger areas protect a greater variety of habitats. Expansive and unbroken wildlands are also imperative for large predators, who need sizeable territories for their livelihoods and who prosper in areas secluded from modern human activities and infrastructures. Wilderness stands as the last bastion against civilization's invasions for agriculture, pasture, resource extraction, hunting, fishing, and settlements. All these developments hinge on the habitat fragmentation that infrastructural sprawl effects via roads, canals, dams, fences, power lines, and the like. Thus, one of the chief goals of rewilding is not only to halt more infrastructural penetration into wildlands, but to undo already existing infrastructures in many natural areas.

Rewilding is the aspiration to set natural areas and processes *free* to express their inherent, creative manifestations. By robustly protecting expansive and unfragmented wilderness, and multiplying rewilding projects around the globe, we can halt the extinction crisis and avert a human-driven mass extinction event. Studies are also increasingly showing that large-scale nature protection and restoration will contribute substantially to mitigating global heating and thus averting many ecological and social disasters in its wake. Importantly, the rewilding vision also moves us toward creating a new existential and ethical foundation in the relationship between humanity and Earth: one built on the virtues of restraint, reciprocity, love, and awe for the flourishing of all life. These virtues are inherent in human nature—indeed recognized and valued by all human cultures and religions—yet they have been severely diminished by the human arrogance that the domination of nature vaunts.

———

To get to the heart of the disjuncture between the new environmentalism's deprecation of wilderness and the aspirational vision of rewilding, I believe we must go deeper into their rift. Opposing stances toward humanity's expansionism constitutes the core difference between the two platforms. Human growth trends include global population increase, expanding economies and trade, rising consumption of food, freshwater, energy, and materials, intensification of industrial animal and crop agriculture, and burgeoning networks of industrial infrastructure. These trends underlie a multidimensional and catastrophic ecological situation, including rapid climate change, global toxification, acidifying and depredated oceans, and extirpations of life-forms, populations, and ecologies that are prefiguring a mass extinction event. New environmentalists acquiesce to the human expansionism underlying these disasters, for growth appears to them as either freighted with human benefits or having too much momentum to challenge. Thus, the new environmentalism prefers to tout solutions within the framework of techno-managerialism: improved management, better governance, more efficiency, as well as technological fixes, breakthroughs, and transitions by which problems will purportedly be solved.

For rewilding advocates, opposing human expansionism is key to healing the natural world and our relationship with the earth. Attempting to work with and around the growth trends will not avert the massive ecological repercussions here and on the way. Ironically, many new environmentalists suspect the same: for example, their receptivity to climate geoengineering and their painting a "bright side" to a mass extinction event appear as implicit confessions that a profoundly impoverished Earth lies ahead.

Accepting civilization's expansionism and hoping at the same time to retain Earth's remaining biodiversity is foolhardy when we consider how biodiversity is faring in a not fully global economy of eight billion people. The aggregated biomass of humans and livestock today dwarfs the biomass of wild vertebrates by 96 percent to 4. This infamous metric starkly captures the planetary consequences of growing human numbers, rising consumption of everything, exploding global trade, and infrastructural sprawl. This being the picture today, what awaits in the near future where the global population is projected to climb to some eleven billion people, who will have (or desire) a middle-class, commodity-saturated lifestyle, and be connected in a more tightly knit global economy? Setting aside what such a world bodes for humanity— hardly looking like "a good Anthropocene"—what, might we extrapolate, awaits the wild nonhuman world?

Rewilding advocates argue that we must not head in that direction to find out, let alone welcome going there. It is vital to degrow and restrain the human enterprise. Embracing limitations involves scaling down our demographic presence, economic activity, and reach of technosphere, while simultaneously generously protecting unfragmented wildlands and rewilding the planet, thus allowing the vibrancy of terrestrial and marine wilderness to flourish again. For the Rewilding movement, holding a positive view of wilderness is paramount in order to fire up the human imagination and engage humanity's emotional intelligence. Our curiosity and love for wild nature and wild beings can inspire us to recoil from surrendering to a human-dominated and human-defined age. In moving toward a rewilded planet, not only can we avert dire ecological and social disasters on the way, we can also aspire to restore the biodiverse and lively planet that Earth inherently is.

8

Picturing the Wild

Ben A. Minteer

HE WANTED to be a pianist. Music was an organizing force in his life starting around the age of twelve, when his father heard him picking out notes on the family piano in their home in the Sea Cliff area of San Francisco. But his hopes met a few stubborn realities. Despite years of study and a promising musical talent, his hands were small and somewhat delicate; they bruised easily when playing more strenuous pieces. He continued to play piano with some seriousness into his twenties, but his interest and pursuit of a life in music eventually faded.

Although possessed of a "wiry frame and considerable stamina," he endured regular bouts of illness as a child—primarily debilitating colds and flu—and remained confined to his bed for long stretches at a time. He was also a hyperactive kid, and his mental health became a concern as he struggled with what he later termed "unsettled periods of weepiness." A doctor prescribed bed rest in a dark room for two hours each afternoon, a course of treatment he resisted, and which, not surprisingly, made things worse. When he contracted measles, he was put to bed for two weeks, with the shades drawn to shield his eyes.

Bedridden again around the age of fourteen, he was given a book that captivated him. *In the Heart of the Sierras* (1886), by J. M. Hutchings, gripped the boy with its exhilarating accounts of "cowboys and Indians" and depictions of spectacular wilderness environs. He devoured every

word, becoming enrapt by the "romance and adventure" of the landscape. He had to see this place. And feel it.

That same year, on June 1, 1916, he visited Yosemite with his family,
riding by train from Oakland and then by open bus from El Portal to
the park. More than half a century later, his description of what he saw
that day retains the sensorial vividness of first contact:

> We finally emerged at Valley View—the splendor of Yosemite burst
> upon us as it *was* glorious. Little clouds were gathering in the sky
> above the granite cliffs, and the mists of Bridal Veil Fall shimmered
> in the sun . . . One wonder after another descended upon us; I recall
> not only the colossal, but the little things: the grasses and ferns, cool
> atriums of the forest.

One day after his family's arrival at the park, his parents gave him a
Kodak Box Brownie, his first camera. He made snapshot after snapshot
during that first visit and would return time and again to Yosemite over
the next few years, sometimes on his own. In 1920, he took a job as custodian of the LeConte Memorial Lodge in Yosemite, the Sierra Club's
headquarters in the park. And his photos gradually became less a documentary record of his excursions in the park and something far more
composed, and more refined. Ansel Adams had found his calling, and
his life's work.

————

Ansel Adams's role in shaping the aesthetic expectations of generations
of viewers toward the American wilderness is undeniable. Forty years
after his death, he remains a ubiquitous presence in the pop culture of
American landscape art, with his prints adorning countless living room
walls, calendars, and day planners, and his slickly designed books con-
spicuously displayed on coffee tables.

At the same time, his development as an artist and conservationist is
little known outside the photographic community. When he is referenced in the contemporary environmental context, it's often an ill-
informed caricature: the photographer who mythologized and, in a

sense, "faked" the wild in his pictures. But Ansel Adams and his work are more important and more interesting than that. And also, more vital.

Adams's photographic approach evolved quickly after that first Yosemite visit, though not always evenly. His early work was pictorialist in character, a soft-focus, "painterly" style that mimicked etching. Beginning in the late 1920s, however, he started exploring the unique technical and visual qualities of photography as an art form, an approach that placed him among other modernist photographers such as Paul Strand, Edward Weston, and Alfred Stieglitz. By the early 1930s Adams had made the shift to the "straight" photographic style with which he'd forever be associated: the creation of sharp prints with expansive depth of field where fine details could be viewed across the image.

The Sierra Club played an outsize role in Adams's life and art from the very beginning. His pictures regularly appeared in the club's *Bulletin* in the 1920s (he was eventually appointed to its editorial board), and he prepared multiple photographic portfolios celebrating the popular Sierra Club Outings in the 1920s and 1930s. In 1934, Adams was elected to the club's board of directors, and his influence within the organization expanded over the decades as his own reputation and visibility grew.

Adams's early work with the Sierra Club also figured in an important series of photographs he made of the area now known as Kings Canyon National Park (fig. 11). A remote part of the High Sierra, the Kings Canyon region had been on the organization's agenda for formal protection; Adams was asked by the club board of directors to travel to Washington, DC, with his Kings Canyon photos in 1936 to lobby for the creation of a national park. His Kings Canyon images were collected in his first book of landscape photography, *Sierra Nevada: The John Muir Trail* (1938), and the volume eventually found its way onto the desk of President Franklin D. Roosevelt, who signed the legislation creating the area as a national park in 1940.

The parklands continued to beckon. In 1941, Adams was commissioned by the US Department of the Interior to create a large mural of the parks for its new building in Washington, DC, an assignment that found him traveling across the western national parks and monuments and taking hundreds of photographs. He was able to think and work in

FIGURE 11. "My unsophisticated presentation of photographs, coupled with appropriately righteous rhetoric, stirred considerable attention in Congress for our cause." Photo: *Middle Fork at Kings River from South Fork of Cartridge Creek, Kings River Canyon (Proposed as a national park), California, 1936*. From Ansel Adams Photographs of National Parks and Monuments, 1941–1942, US National Archives.

a large, panoramic format, a context that allowed him to draw upon his technical and artistic strengths in image making, including his process of "pre-visualizing" the photograph he intended to produce by syncing the image in his head (as well as its desired impact on his audience) with his manipulation of photographic and printmaking processes. Many of the pictures Adams took as part of this project have entered the visual lexicon of Western landscape photography and national park history, including arresting images of Yosemite, the Grand Canyon, and Canyon de Chelly.

One of the most striking series of photographs from his Department of the Interior work is a set of images of a storm over Wyoming's Grand Teton National Park (fig. 12). The dramatic and awe-inspiring qualities that Adams was able to render in such images, his "operatic" style, as the photographic historian and curator Rebecca Senf describes it, convey the power and glory of the American wilderness, bathed in a highly romantic and often nationalistic light. Indeed, despite his modernist sensibilities, photos like *The Tetons and the Snake River* inevitably conjured an older notion of the natural sublime, an aesthetic motif extolling the awe and forcefulness of wild nature, that animated landscape painters like Thomas Cole and Albert Bierstadt in the nineteenth century. As Senf notes, the lack of much immediate foreground in Adams's *Tetons* photos, that is, where we might be standing as the viewer, confers an omniscient observer effect to the images, creating a scene that comports more with an emotional and aesthetic reality than any empirical one. Photographs like this are the reason why Adams was often referred to (by both admirers and critics) as "John Muir with a camera."

And in fact, on those occasions when Adams articulated his wilderness philosophy, he often struck a Muir-like tone, commending the sanctity of the wild landscapes toward which he turned his lens. As he wrote in his 1961 essay, "The Artist and the Ideals of Wilderness":

I believe we must agree that wilderness is more than physical appearance, more than decoration, more than an arena for sport, and more than a proving ground for physical prowess. It must stand as a symbol

FIGURE 12. "You must have quietness and a certain amount of solitude. You must be able to touch the living rock, drink the pure waters, scan the great vistas, sleep under the stars and awaken to the cool dawn wind. Such experiences are the heritage of all people." Photo: *"The Tetons and the Snake River," Grand Teton National Park, Wyoming, 1942.* From Ansel Adams Photographs of National Parks and Monuments, 1941–1942, US National Archives.

of qualities beyond the structure of routine life. It competes with no religion; rather, it suggests a new religion, the revelation of which is comprehension of the vast cosmos and the ultimate purpose and validity of life. In this sense, wilderness is "sacred" and the wanton disturbance of a twig or a stone or the casual murder of an animal— these constitute violations of the temple.

Muir, who in the early twentieth century railed against the "temple destroyers" that sought to dam a valley in Yosemite for a water project, could easily have penned these words (though he would have dropped the scare quotes bracketing "sacred"). And yet as we'll see below, Adams would elsewhere (and later) soften such views, albeit modestly.

Still, there is no doubt that for most of his life Adams hewed to a hard-charging position on park preservation, one that allied him, at least for a time, with his friend and fellow Sierra Club leader, David Brower. That included Adams's anger at the Park Service's catering to recreation and mass tourism interests in the 1950s and 1960s, when the new Interstate Highway System and a rising standard of living converged to juice park visitation. For conservationists like Adams and Brower, it meant that the wilderness values of the parks, especially a sense of solitude and an aesthetic experience outside of what they saw as crass commercial manipulation, were in peril of being destroyed by growing hordes of "industrial tourists," to evoke Edward Abbey's memorable derogation.

The popular influence of Adams's images and the Sierra Club's larger conservation vision reached a crescendo in 1960 with the publication of *This Is the American Earth*. Edited by Brower and pairing the photographer's images of western wildlands with text by the photography critic Nancy Newhall, the book drew from an earlier exhibition of Adams's photos at the LeConte Lodge in Yosemite in the mid-1950s (he'd clearly come a long way from lodge custodian). An image from Adams's Tetons and the Snake River series prominently appears in the volume alongside a selection of his photos of Yosemite, Glacier National Park, and an array of other heroic western vistas.

But the book also carried a stark warning, patent in the inclusion of incongruously paired photos of a smog-filled and crowded postwar Los Angeles by the photographer William Garnett. If we didn't get control of ourselves, the pictures suggested, the wild places on the American landscape could suffer a similar fate. As conservationist and Adams's collaborator William Turnage writes, the book in hindsight ushered in the era of popular environmentalism, arriving a couple of years as it did before Rachel Carson's *Silent Spring*. It no doubt also played a role in helping Brower and the Sierra Club influence public and political opinion in the run-up to the passage of the Wilderness Act in 1964.

The success of *This Is the American Earth* inspired the Sierra Club's aptly named "Exhibit Format" books, a series of large format, lavishly illustrated, and expensively produced volumes. Among them were two books by Adams's fellow photographer and Sierra Club board member

FIGURE 13. "On the assumption that all art represents about the same thing—the spiritual penetration of reality, the creation of configurations of meaning out of chaos, the revelation of beauty and the identification of man with his environment—I must recognize my own medium of expression as but one way of seeing and saying." Photograph taken by J. Malcom Greany. US National Park Service, Yosemite Historic Photo Collection.

Eliot Porter, including *In Wildness Is the Preservation of the World* (a copy of which Brower gave to President Lyndon B. Johnson at the signing of the 1964 Wilderness Act), and *The Place No One Knew* (1963), a photographic elegy for the recently flooded Glen Canyon. Some of the books were tied to particular campaigns for preservation in the face of proposed development or resource extraction; others were pitched at the general level of environmental consciousness raising about the aesthetic and cultural value of the wilderness landscape in late twentieth-century American life.

As historian Finis Dunaway notes, Brower saw the books as art objects with the potential to reach policymakers and what today would be branded "influencers" in the effort to preserve the American wilderness. Expensive volumes, they were also tokens of conspicuous consumption. Writer John McPhee, whose 1971 book on Brower, *Encounters with the Archdruid*, helped to cement the preservationist cause in the minds of a generation of readers, relayed that Brower himself was surprised to find that people were willing to shell out that much for beauty. Yet the books sold, and also helped to boost the club's membership during their heyday in the 1960s. By the end of the decade, the series had grown to nearly twenty books; and yet the expense of their production and Brower's domineering management style ultimately led to his undoing with the organization. Beauty, apparently, did have its fiscal limits, at least in the club's publishing program.

Adams clearly understood the power of the Sierra Club books to disseminate his vision and advance the campaign for wilderness preservation. Less political than Brower, Adams assessed their potential from the artist's standpoint, and in the aesthetic response of the public. As he saw it,

> photography, in conjunction with the appropriate written word and suitable media for distribution, may well be the art which can effect the miracle—reaching deepest into minds and hearts and strengthening understanding, elevating the spirit, and bringing the certain confidence necessary to support the sense of mission which our people must embrace or approach disaster . . . I feel that—without

reservation—unless some great spiritual experience is evoked, some deep excitement and sense of purpose stimulated within our people, our cause is lost.

Environmental historian Roderick Nash, whose book *Wilderness and the American Mind* remains the most important study of the idea of wilderness and the development of the conservation movement, suggested that Adams held a strongly biocentric view toward wild nature, one that put the preservation of naturalness over all else. While Nash was clearly approving of Adams's preservationist zeal, many saw it differently.

The art historian and photographer Deborah Bright, for example, found Adams's exclusivist taste for a purist wilderness vision deeply troubling. His "aesthetic contempt for mass culture," she wrote, was "palpable." Other scholars were even more cutting, especially the eminent essayist and cultural geographer J. B. Jackson, who had little use for Adams's imagery and the books that helped put it into wide circulation. For Jackson, whose somewhat contrarian admiration for roads and the "altered" landscape led him to be suspicious of traditional nature preservationism, the Sierra Club's publication efforts had resulted in a series of "anti-urban, antitechnological, antipeople, antihistory books and pamphlets, all anthrophobic, all urging us to worship nature."

Such criticisms dovetailed with a broader unease about the American wilderness ideal—or "wilderness myth" that emerged in force around the mid-1990s, most notably in historian William Cronon's provocative critique, "The Trouble with Wilderness." Arguing that our notion of wilderness was a cultural creation rather than an objective mirror of nature, Cronon suggested that environmentalists needed to devote more attention instead to the places in which most people lived, worked, and played. And he worried that the uncritical attachment to a historical and ecological phantom, that is to say, the ideal of the pristine wilderness, would lead us to devalue the more prosaic parts of the landscape critical to achieving long-term sustainability on the planet. The wilderness tradition in American environmentalism, in other words, elevated an imagined, Edenic wild over the real, lived-in places that always seemed to fall outside Adams's frame.

Adams has also been singled out by the journalist Mark Dowie, who takes him (and his nineteenth-century predecessors like Carleton Watkins) to task for creating visual fictions in places like Yosemite. Their photos, he writes, are "completely bereft of humanity or any sign of it having been there. Here they said (and they all knew better) is an untrammeled landscape, virgin and pristine, not a bootprint to be seen, not a hogan or teepee in sight." Such aesthetic choices were clearly deliberate on Adams's part. As the writer Rebecca Solnit once observed when accompanying photographer Mark Klett in his rephotography of Adams's and others' Yosemite views, some of Ansel's iconic "wilderness" images of the park were taken very close to the road but were cropped and facing away from development "to make the place look pristine."

———

There is little doubt, then, that Adams sought to convey an image of the American national parks and wilderness that was uplifting, pure, and timeless in its beauty and environmental character, as well as one that offered a stark visual contrast with the rest of the landscape with its highways, strip malls, and subdivisions. It was an artistic vision and style honed through years of technical innovation and picture making and driven by the photographer's own emotional and aesthetic response to the monumental wilderness landscapes of the West. Not to mention his desire to evoke that same response in viewers—"reaching deepest into minds and hearts" to inspire a passion for nature appreciation and preservation. And yet as the great writer and conservationist (and Sierra Club colleague) Wallace Stegner argued, viewing Adams as just a romantic nature photographer sells him short.

Although legendary for his wilderness panoramas, Adams made tens of thousands of photographs in his lifetime, demonstrating a more versatile repertoire of views and subjects than is typically known. That includes smaller-scale close-ups of rocks, tree stumps, bird nests, and even blades of grass, indicating the photographer's abiding interest in the landscape's humbler natural features as well its more titanic forms. "The appreciation of the natural scene should not be limited only to the grandeurs; the intimate details and qualities of our immediate environment can be revealed and appreciated," Adams remarked in his 1961 commencement address at Occidental College.

He also took many portraits, including of Native American subjects, albeit not showing them living in Yosemite and other classically "wild" places. Like most conservationists of his era, Adams didn't reckon with the difficult and painful cultural histories of the wilderness locations that caught his eye and stoked his imagination. As an artist, he made his choices about what he wanted to present, and what he wanted us to see, as all artists do. Still, for someone with such a wide field of vision he was undeniably myopic when it came to the Native American presence on landscapes like Yosemite.

Adams's assignment for the Department of the Interior in the 1940s also found him photographing highly altered, industrialized features such as Boulder Dam (fig. 14)—conveying a technological rather than naturalistic expression of the sublime. And to pay the bills, he did fairly extensive commercial photography early in his career for IBM, AT&T, and *Life* magazine. Although much of this work is not nearly as distinguished or distinctive as his more well-known wilderness imagery (even Stegner had to admit that his photographs of people were comparatively "less successful"), it's a reminder that Adams was a working artist with a range of interests and experiences, not simply an avatar of puritanical wilderness preservation.

His environmental advocacy within the Sierra Club, especially later in his life, was also more nuanced than is often assumed. A revealing example is Adams's role in the Diablo Canyon controversy in the late 1960s, an issue that caused significant strife within the Sierra Club leadership.

In the early 1960s, California's Pacific Gas & Electric Company (PG&E) purchased land in the Nipomo Sand Dunes overlooking the Pacific Ocean south of San Luis Obispo, with plans to build one or more nuclear power plants. The Sierra Club, including Adams, had campaigned to move the project away from what was seen as a geologically unique landscape to an alternative and arguably less fragile site, Diablo Canyon,

FIGURE 14. A far from pristine landscape shot by Adams as part of his Department of Interior assignment in the early 1940s. *Photograph of the Boulder Dam from Across the Colorado River, 1941.* From Ansel Adams Photographs of National Parks and Monuments, 1941–1942, US National Archives.

to the north. PG&E agreed to the plan, but Brower, no doubt still traumatized by his perceived "sacrifice" of Glen Canyon not long before, ended up revolting against the agreement, and pulled some of the club's board members into his corner. It led to the proposal being put to a full vote by the club's membership, the first time a conservation issue had been voted on in this fashion. Brower lost; the majority of club members and rest of the board, including Adams, overwhelmingly supported the Diablo Canyon agreement. The episode clearly exacerbated a personal breach that had been growing for years between the two old friends. Brower was fired by the club's board in 1969 and Adams resigned in 1971. A plant with two reactors was eventually built and became operational in the 1980s.

Originally slated to be decommissioned beginning in 2024, the plant's life has recently been extended for another five years in the wake of energy reliability concerns—and as part of a larger renewable energy transition plan in the state.

Adams's thoughts on the Diablo case are interesting, not least because of the pragmatic position he took on energy development and the realities of land use decision-making within the cause of conservation advocacy. As he wrote in a 1969 letter about the Diablo dustup to the *Toiyabe Tatler*, the Reno, Nevada, Sierra Club newsletter:

> There are hundreds of areas along the coast equal to or surpassing Diablo Canyon; I wish we could save ALL of them, but the realities of life prevent . . . As I have said often, *Cooperation is NOT Compromise*. It is fundamentally wrong to approach the problem by assuming that the government, highway, and utility groups are "enemies." This is a belligerent and somewhat paranoid attitude which many conservationists assume (I felt that way once, myself!). It gets us nowhere. It removes the possibility of rational discourse and cooperation.

Despite his penchant for moralizing and self-described "righteous" anger over the threats to his beloved parks and wilderness, Adams's preservationism seems to have become at least partially tempered as the years wore on. In the Diablo dispute, some of this can likely be chalked up to the difference in context—the beautiful but comparatively common landscape of the oak woodlands and chapparal canyon of Diablo was no Yosemite or even Kings Canyon. But Adams was also enough of a realist about the human project, including the development of nuclear power, which he believed would be less polluting of the special places he cared so deeply about. Although he made it clear he wasn't advocating an ethos of "giving in," cases like Diablo were a reminder to him that hard choices had to be made, and social and political realities acknowledged. As he lamented in that same 1969 letter, "I wish all members of the Sierra Club could realize they live in a different world than that of John Muir!"

———

Adams's pictures show us a view of nature in its ideal form, a font of values defined by their contrast with the highly worn and humanized landscapes within which most of us live and move about day-to-day. His magnificent images of the American parks and wildlands were direct efforts to create an emotional and ethical response in the viewing public to the burnished wilderness tableaus he created. And yet Adams also knew that his was not the only way to see and think about the wild, or about nature and the landscape more generally. It was one (shimmering) tile in a larger cultural and environmental mosaic.

It's also important to remember that Adams's photos are, first and foremost, artistic creations. He wasn't a documentary photographer, and he deliberately chose not to be one. He was an artist, and also a conservationist who sought to present and instill a sense of spiritual, therapeutic, and psychological fulfillment in the parks and wilderness lands of the American West.

In that sense we can think of him as working with what the iconoclastic filmmaker Werner Herzog calls "ecstatic truth" rather than documentary realism. His goal was not to "transcribe" nature as it really is (assuming that would even be possible), but to use photographic art and his mastery of technique to produce what he saw as a higher, that is, an emotional and spiritual, truth about the wilderness and our relationship to it. Just as Edward Abbey laced fictional elements into *Desert Solitaire*, which nevertheless maintains the appearance of a journalistic account of his time in the sunblasted parklands of the Southwest, Adams created stylized mashups of experienced reality and artistic vision in his photos, all in the service of a highly disciplined artistic and conservation ethos. In doing so, he taught generations *how* to see and photograph natural scenes and the wilderness, and above all how to appreciate them.

Despite the popular and often well-worn familiarity of his photography, Adams's faith in the unique power of the wilderness image and experience remains vital a century after his early photos of the Sierras graced the pages of the *Bulletin*. His work and vision have an important place in our environmental ethics today, especially as we qualify, revise, and diversify the environmentalist project in a time of often dizzying social and ecological change. We still need his pictures, even in these human-centric times, if only because they remind us that we haven't grown "too big for nature," as some of the more exuberant boosters of the Anthropocene have asserted.

Rod Nash, though, didn't have it quite right. Adams wasn't a true biocentrist, at least if that means he sought to sanctify places like Yosemite for their own sake outside of any experience of them. Instead, it was about what his carefully composed wilderness views meant for a society grappling with its own growth and development, and what they meant (and still mean) to us as people.

And of course, it was about what they also meant to him. It's hard not to think of Adams as spending a lifetime chasing that original sense of wonder and exhilaration he felt during that initial trip to Yosemite with his parents, a deeply personal attempt to recapture the transformative power of the sublime park and wilderness lands of the West. It's no small irony that the boy forced to convalesce in the dark ultimately became an internationally celebrated artist famous for his command of photographic exposure. And it makes a recollection written near the end of his life of seeing Yosemite that first time especially revealing, even poignant in hindsight: "There was light everywhere!"

9

In Feral Land Is the Preservation of the World

Kathleen Dean Moore

IT WAS just eleven acres of failed wheat fields grown up to tall fescue and field daisies, bordered by a stream that flowed through a dense ash swale in western Oregon. Some decades ago, an economics professor named Floyd thought he could make a killing raising wheat on those acres, taking advantage of grain shortages and starvation in the former Soviet Union. That plot fizzled: maybe the land didn't give itself up to wheat, or maybe the Russian people didn't suffer enough to make the undertaking profitable. So he let the land go—he let it go—and it became a feral field.

The land was feral the way a cat is feral when it is thrown out of a Chevy that scarcely slows down—that frightened and fierce, that opportunistic and desperate. The cat will do what it has to do, remembering in its soft belly how it felt to be fed, remembering in its jaw muscles how it felt to be wild, maybe surprised that its haunches know how to spring after a sparrow, surely relieved to taste the sharp little bones and bitter muscles, to tease the feathers out of its teeth.

The stream remembered the path through the trees, fleeing ditches and mires ahead of high water. Camas bulbs remembered the warmth of new light. In their terrible hunger, blackberries and fescue crawled over the countryside, pillaging. The seeds of tarweed found again the strength to force strong sprouts between pebbles. Flocks of goldfinches fed upside-down and sideways and every which way, rattling dry teasels. The fields were disheveled. They looked like no one cared for them. But what was beautiful about the view across this land was the striving. No one who saw this land would be able to deny the life force, the surge of continuing. Shattering seeds, insistent tendrils, screeching jays.

Floyd offered the land at a going-out-of-business price, which was $24,000—just about the amount bequeathed to me by my Aunt Eva, who was a librarian's assistant. My husband and I bought the land.

"What are you going to do with it?" friends asked us.

"Nothing," we said. "Visit it. Share it with our grandchildren, if we are lucky enough to have any."

"What is your management plan?" the ag extension agents wanted to know.

"Leave it alone."

Confused, the agents left *us* alone, but the tax assessors were having none of this. They insisted that we turn the land to some human purpose and profit, still faithful to the English philosopher John Locke's theory of property: by mixing his labor with the land, Locke said, a man gives value to something that was worthless, and so he has a rightful claim to the land-value he has created. To avoid punitive taxes, we planted 221 Douglas fir seedlings on the north rim of the field and called it a tree farm. That was arguably true—we did grow trees. How can you keep a Douglas fir from growing? But our secret plan was to keep the place safe while it got to its feet again. We would take in the wounded land. We would care for it— clumsy but well-meaning—until it could care for itself, and then we would set it free forever. In our minds, these riverbottom old fields should go wild, lively with frog song, crawling with newts, alive to their own purposes and lifeways. We called it The Farm, but that was a disguise.

————

Feral, adj., "in a wild state, as an animal may become feral when it escapes from captivity or domestication," or is driven out or abandoned by the people who once kept it. Feral *land* is—by analogy—land that once was cultivated or in another way turned to human purposes, but has been abandoned and allowed to find its own way.

Depending on whom you ask, humanity has already domesticated or "significantly altered" 75 percent of the land and 66 percent of the ocean. "Domesticated" is, of course, a euphemism. Turning land to domestic purposes has so far been marked by violence and violation, killing or driving out the denizens of forest and marsh, as the land is scraped, paved, plowed, poisoned, and in other ways laid waste. As a consequence, since 1900, the world has lost 20 percent of its native species, 40 percent of its amphibian species, 33 percent of reef-creating corals, and 33 percent of marine mammal species—a rate of extinction perhaps one thousand to ten thousand times the background rate.

This planetary disaster is the unsurprising consequence of what philosopher Eileen Crist calls "entitled instrumentalism." This is the powerful, wide-spread, seldom-challenged assumption, central to colonial expansion and extraction, that the entire planet was given to some portion of humankind to use for their own ends. On this model, private land, public land, the commons, marine reserves—it makes no difference—all are presumed to be owned by humans, individually or collectively. On the other hand, wolves, mangrove swamps, glaciers, kittiwakes have no property interest in their places. The result? "The wholesale assimilation of the natural world into the human-owned instrumental domain," and the consequent destruction of lives approaching the magnitude of destruction caused by the impact of the Chicxulub asteroid that wiped out an estimated 76 percent of the species then on Earth. That was the fifth extinction. This one, the extinction currently underway, is on track to become the sixth.

It's going to take some heavy lifting to challenge the claimed right to, frankly, *everything on Earth*. But this is what biodiversity scientist E. O. Wilson says the world must do to stop the sixth extinction. The goal of his Half-Earth Project is to protect fully half the world's land and seas from anthropogenic harm. If we can manage to do that, he calculates, 85 percent of Earth's species can survive the onslaught of expansionism. At that point, he says, "Life on Earth enters a safe zone." *A safe zone*. Imagine:

half the earth free of human exploitation and extraction, land and seas devoted to plants and animals living independently or in harmonious relations with humans, as they have done before and may do again. This is the beating heart of the wild.

————

The Farm went its own way for a decade. Sure, my husband and I gave it a couple of nudges. We dug out blackberry canes. Planted a couple of oaks. Nailed up a shabby suburb of wren houses, owl houses, bluebird houses. Invited local land conservationists to tear out patches of fescue and, in their place, plant Kinkaid's lupine, hosts to the endangered Fender's blue butterfly found in neighboring fields. And we did visit the land, often with our two young grandsons. Imagine small boys and swallows careening over the fields, catching bugs. Imagine garter snakes licking the smell of shadows. Imagine the joyful astonishment, when the boys climbed the limbs on a fifty-foot Douglas fir and found the nest of an owl.

We dug a pond, right there at the edge of the swale. It was a vernal pond, designed to fill with water in the fall and winter, when waves of rain clouds pounded the Oregon coast. All through the spring, showers refreshed the pond. In summer, the rains stopped and the pond dried up, done for that year, ready for the next. The pond was not very big, but big enough for cattails, where redwing blackbirds shouted. *Okaleeee*. And big enough for mergansers wooing their mates. *Twang*. Pacific tree frogs came in huge numbers, singing their hearts out. *Ribbit*. Red-legged frogs sang underwater. *Uh-uh-uh-uh-uh*. In the damp spring, rough-skinned newts plodded to the pond, shimmied in, and began to court, a behavior called amplexus, when a male holds a female in an embrace that lasts for hours, as she swims around the pond, lugging her lover piggyback. Days or weeks later came eggs, laid one by one, and then the detritus of the pond was alive with the tiniest imaginable newts, each wearing feathery epaulettes of gills.

"I found a newt," a child called out. His little brother splashed over in high boots already filled with water. "Don't touch. Poison glands in the skin," the older boy solemnly advised, and the little pointer-finger retracted.

————

If you were to ask me where the planet will find land to devote to nature's half of the earth, I might think first of legally protected wilderness reserves. Say, the expanses of Pacific Northwest old-growth forests in the roadless areas of Alaska's Tongass—hemlocks chest-deep in ferns and devil's club. I might think of fourteen-thousand-foot peaks in the Flat Tops Wilderness Area, protected by coveted backcountry hiking permits. I might think of Biosphere Reserves in Brazil's Pantanal wetlands, a place of hoots, taps, screeches, howls, flapping wings, and falling water—bordered by drained fields of grazing cattle. I might think of Antarctica, wild and cold, its edges breaking off. Or a red-rock canyon wall in the remnants of the Bears Ears National Monument, its history traced by the shadow of a condor's flight across layers of sandstone and years of legislation and court battles. Yes, of course, we must defend these embattled formal reserves with all we've got.

Beyond designated wilderness areas, there are wild places to defend, what we might call "undeveloped" land, which is to say, not yet fully exploited. But wild places are few and diminishing, pushed back primarily by the expansion of agriculture and mining. In fact, whether there is anything wild left on the planet is now a matter of definition; if "wild" means unaffected by human hegemony, then as Bill McKibben pointed out decades ago, there is no wildness left.

No. Wildness is not enough for the preservation of the world. To get half the world under protection, global and local governments will have to turn to feral land, setting vast tracts of land free again—land once thoroughly trampled and trammeled, then left to itself. This is how the world will preserve the natural forces of healing that may yet save planetary life. In such feral land, I believe, is the preservation of the world. Not in Thoreau's wildness—dangerous, untamed Mount Katahdin, swirling with icy wind. But rather in land that is not pure; call it ruined and in recovery. Not isolated; but in a mutually healing relation with people.

We can choose to return cornfields to the larks, return cattle ranges to the bison, return suburbs to the forests, return quarries to the skinks. I imagine this will be a matter of economic incentives at first. Later, one can imagine the blossoming of an Ethos of Return. What was taken can be given back. What was ruined can sometimes be healed. What was driven out can be invited to return.

An Ethos of Return begins by renouncing the claim of human entitlement to use all of creation for human ends. By what right, one might well ask, do humans take it all? I can't think of any good answer. One might argue that God gave Creation to human beings, but why would He? What reasonable Being would specially create the exquisite spring chorus of frog song, then invite humanoids to bulldoze the marsh, destroying the frogs and their songs? Surely, on these terms, this taking, this taking, this endless taking, is a theft of the sacred—literally a sacrilege.

Once enough people renounce the human claim to own it all, the door opens to a moral world defined by sharing, gratitude, and restorative justice. Sharing: taking only a portion of what is offered, so that others can have their portion too. Gratitude: recognizing abundance and sustenance as gifts, not bills payable. Gratitude requires honoring the gifts, caring for them, and nurturing the earth's capacity to give such blessings. Restorative justice: returning what was unfairly taken, and so coming back into a right relationship with the ones who were wronged.

The analogy with restorative justice for victims of mortal crimes and moral insult, including Indigenous people and descendants of enslaved people is strong; if you or your ancestors have put yourselves in a favorable position by taking land and liberty from others, the way to restore a just relation is by returning what was taken. There is nothing simple about recompense, and the process will be contentious and maybe endless; but the moral principle is clear. What was unjustly taken should be restored. So it is with the beings of land and sea whose habitats have been taken and often destroyed—the birds of the air, the fishes of the seas, every living thing that moves upon the earth.

An Ethos of Return is a virtue ethic. With a heavy hand, essayist Brian Doyle identifies the venality of unsubstantiated claims to all the earth. "And what kind of greedy criminal thug thieves would we be as a people and a species if we didn't spend every iota of our cash and creativity to protect and preserve a world [that holds the rough-skinned newt]?" But ask the question the other way: How can humanity achieve a higher level of moral development by relinquishing its devastating control over the earth and cultivating instead a relation of mutual thriving with the life-giving rivers, plants, and animals with whom we share the world? What can the scrappy independence and sustaining interdependence of lives on feral land teach us about living in a gasping world that we have strangled with our own hands? What can we learn about an Ethos of Return by watching new cooperative agreements unfold, where Indigenous lands are returned to their original inhabitants for conservation and preservation based on traditions of mutual thriving—a nascent attempt at redemption by return. And so perhaps people can begin to redeem themselves by setting captive land free. This is the open heart of the wild.

———

Then came the summers of 2019, 2020, 2021, the years of drought and heat at The Farm. First, some Douglas firs began to die. They withered from the top down, dropping needles from the high branches, which became bedraggled and empty of everything but cones. Then imagine needles yellowing, the yellow descending, until the entire tree dropped its needles and stood naked—a sad, obscene striptease. Two of the oaks we had planted died. The ash in the swale stood on hard conglomerates of leaves and dried mud. The river sank into its stones.

And then, under the hard stare of the sun, the pond began to shrink. Too soon. Too soon for the tadpoles. Too soon for the larval newts. In a damp year, the froglets would grow their sprongy legs and hop into the damp, dark swale before the pond sank into pondweeds. In a damp year, the larval newts would resorb their tails, no longer blade-like, but oval in cross-section. Their bones would harden and their skin thicken. They would resorb their gills and develop lungs, and then they would crawl out of the pond into the moist refuges underground or in a rotten log, to spend a couple of years before they matured enough to come back to the pond to breed.

FIGURE 15. The author's grandchildren and their father and grandfather search for any sign of frogs and newts in the drying bed of the pond at The Farm. Photo by the author.

But these were not damp years, but years of record-obliterating heat. Our grandsons waded into the warm pond water in their tennis shoes, laces floating, socks sagging. They scooped nets through the waters, gathering what little lives they could find. "Hurry," they directed us, and emptied their nets into buckets, emptied their buckets into the stream, came back for more. All day, they rescued small beings from the hardening mud. But soon only puddles remained of the pond, black with wriggling lives. "Hurry. There's no time," and it was true. There was no time left.

When the boys returned to the pond a week later, there was no water. There were no froglets and no lacy salamander larvae. All that remained was a black, tarry slick where the pond had been—the dried bodies of soft beings who would never sing, never hug piggyback, never lay eyeball eggs on pondweeds. By then, our grandsons were too old to let themselves cry.

———

Tragically, the effort to return land to a feral condition may be helped immeasurably by the fact that global warming will drive people off large tracts of land, made uninhabitable. These abandoned lands will become feral not because humans release the land to its wild state, but because humans—driven away by heat, storms, drought, floods— abandon them to their fates. There will be vast stretches of newly feral lands: Farmers' fields abandoned as salt water inundates the plains. The desert sands after the oasis springs dry up. Southwestern cities that imagined water when there was none; first the golf greens, then the cotton fields, then the suburbs, then the city itself, emptied and left to the lizards. Mountain forests, burned to bedrock and abandoned before they burned again. Coastal villages, left to crumble in giant waves no longer restrained by sea ice. Cities, too hot to inhabit. Global warming will force people to flee, leaving the land to live without them as best it can, and who can predict if that is for better or for worse.

But clearly worse, there is yet another source of feral land. Yes, global warming will indirectly create zones of land gone wild. But human avarice is creating them directly. Maybe you call them sacrifice zones, maybe you call them tailings or superfund sites or leach beds or well pads or strip mines or mountaintop removal. Corporate decision-makers create places where no humans can live and abandon them: gasping, poisoned, leghold-trapped, feral lands, thrown off the speeding Chevy. This is the broken heart of the wild.

———

Leggy now, and strong, the boys are limbing up the Douglas firs at The Farm, preparing for fire season, which now starts in the spring in Oregon and extends to November rains. It's hot, stickery work, but they hack away with hand-saws, removing the lower branches that might allow a grass fire to climb into the crowns of the trees. They pile the branches on the gravel lane, where their parents load them onto a trailer and deliver them to a site where the County will tear them into wood chips and haul them away.

Our family has no doubt the fields will burn one of these hot, dry years. They have burned regularly over the centuries—surely ignited by the explosion of Mount Mazama seven thousand years ago, then burned by the Calapooya people to create green grazing brush for deer and open savannahs for the spreading oaks. Our work is to keep the fire in the grass, preventing it from crossing the lane and burning the neighbors' herbicide-dried vineyards and chateaux faux.

The pond's bowl will burn too. A lot of fuel has built up there and toasted to tinder—papery cattail leaves and pennyroyal whose dried leaves smell like mint. Ninety degrees, a hundred degrees, a hundred and ten. Then even the place of frog song and newt sex will go up in flames. The boys hope to save the owl's nest and imagine that maybe the snakes can save themselves, down in their maze of holes. They are excited about what else will grow up with the green sprouts of new grass, and curious about the underground bulbs of blue camas, waiting for the signal fire.

———

Feral land throbs with the energy of pure possibility, driven by a trembling urgency toward ongoing life. When the heavy human hand lifts off the land, who can predict what plants and animals will do? What survives, what dies, what thrives, what rapidly transforms into something never imagined.

Feral land has a memory of what has been and may yet be again, patterns of swallowing and being swallowed. In feral lands unleashed from extractive, exploitive purposes and plows, cowering life-forms can stagger toward renewal or burst into new abundance—or make way for something new. This last is important: Feral land snarls with the urgency of super-powered evolution. Here is where new forms of life can root or take wing.

And here is where a new relation between human and wild can begin. After centuries of expansion, scarcely restrained by law or morality, we can begin to set free what we have enchained. No one says this sort of cultural about-turn will be easy. No one says it is even possible. But it is necessary. And it has to happen quickly. Releasing land—letting it go—is a form of repentance: Great numbers of us have taken what does not belong to us and have done it great harm, at terrible moral and existential peril to ourselves and others. Now, knowing better, we can release it from our grasp and let it go free. This is the redemptive heart of the wild.

PART III

Knowing Nature in the Human Age

10

Revealing and Reveling in the Story of Nature

Thomas Lowe Fleischner

I WALK with my wife, Edie, on the south side of Granite Mountain, the peak that looms above our northern Arizona town, at the base of which both of our children were born. At the trailhead, rain droplets begin to spatter and the sky rapidly darkens, as the sound of thunder grows closer and more frequent. We head out through the ponderosa pine forest, reveling in the feel of damp ground under our feet—this landscape has endured two years of drought, and the possibility of fire has threaded through our conscious and subconscious mind for the past several months. But this year the summer monsoon—localized thunderstorms seeded by subtropical moisture pulled up from Mexico—has blessedly materialized. Life feels bountiful at every turn, and colors glow vibrantly, in spite of the gray sky. Abundant creamy white flowers of mountain mahogany, a wild rose, sweeten the air. An occasional scarlet *Penstemon* brightens the forest floor.

Then, the day's highlight. Edie notices, just a few feet off the trail, a miniature snake curled into a tight circle, just three or four inches across. Bold brown patches along its back contrast with a creamy background, and strong black marks cross its petite face and extend back toward its tail. We are so staggered by this compact and unexpected beauty that our analytical mode—*Who is it? What is its name?!*—doesn't fully kick into gear. What is clear, though, is that this diminutive being is a very new life, most likely on Earth for only a couple of weeks. And that it appears perfect in its tight coil, nestled precisely within the curve of a fallen branch. Truly, this feels like a gift. We are stunned, and grateful that we are *here*, at the right moment to bear witness to this momentary miracle. The slick-smooth snake remains completely motionless, except for opening its tiny eye when I crouch to take a photo. After two or three minutes, we back away, so as to leave her or him in peace. Thunder sweeps back in, adding a touch of menace to the atmosphere, and our pace quickens.

As we descend toward the bottom of the valley, we hear the thrilling sound of moving water. A mountain stream now flows, creating small riffles over chunks of granite. This is, in itself, astounding, because this had been a bone-dry corridor of coarse granitic sand for most of the previous year. To have mist in the air, small cloud tufts along high ridges, as in a Chinese watercolor, to hear the omnipresent sound of moving water—all this is the greatest of offerings. Along this stream corridor broad-leaved trees glisten with beads of rainwater. From a willow, our attention is drawn by the effervescent song of a lazuli bunting. Tracking the sound, we finally find the bird and witness its brilliant blue (lazuli!) head as it flies a football-field distance to sing again. Nearby, we find one of our favorite late summer flowers—one we seem to forget each winter: long-flowered four o'clock, whose scientific name, *Mirabilis longiflora*, does a fine job of describing both its morphology and its specialness. Its narrow floral tube is longer than my hand, flaring at the tip into an off-white bell, with bright purple stamens dangling beyond. All, when one bothers to look, quite extraordinary. As is the solitary bold white flower of sacred datura, and the up-and-down song of a blue grosbeak. Only later, upon return home, do we dig in to a serious effort to identify the snake, with the help of a couple reptile field guides, each of which represents a compendium of countless years of work by many people. We confirm that our snake was a juvenile Arizona black rattlesnake—the young of which look completely different than the dark adults, and which is nearly endemic to this state. This knowledge, too—a gift from naturalists of the past and present—adds to the sense of beauty, mystery, and gift that the original sighting prompted.

Any morning, any place, any person. I believe my homeland here in the Mogollon Highlands—the vast upland connecting land that stretches across Arizona into New Mexico, where species converge from deserts to the south, plateaus and mountains to the north, and the Great Plains to the east—to be extra special, but in truth, every place is. Each landscape harbors miracles and mysteries. Natural history is the practice of being there, paying attention, and being receptive to revelation. It is the practice of falling in love with the world.

Telling the Story of Nature

Humans are made to do natural history. I mean this quite literally. Human consciousness was forged in natural history's workshop—our patterns of attention sharpened as we watched for danger and searched for food. Our primary senses—seeing, hearing—served both functions: finding food and avoiding becoming food. Our cleverness and knack for language and storytelling allowed us to create culture, and thus to pass down knowledge immediately, person-to-person. Thus, our ecological adaptations sped up, as we were partially liberated from the much slower pace of organic evolution. The necessity of identifying species, tracks, patterns of animal behavior, edible plants, and ecological niches shaped our perceptual and cognitive capacities. Human psyche—how we sense the world, how we interact with one another, how we think and feel—represents the product of our evolutionary responses to the value of attentiveness in our earliest predator-prey interactions.

For millennia, human survival depended upon the attentive practice of natural history. Living well—indeed, living at all—depended on knowing which animals tasted good, and which ones could hurt you. On when and where they could be encountered. On which parts of which plants provided sustenance, which provided the raw materials for tools, and which made you sick. On where water could be found.

Should one doubt that attentiveness and careful observation—the core of natural history—are our evolutionary legacy, just watch any small child, anywhere in the world. They will crouch to peer at an insect crawl-ing past; turn over stones to see what lives beneath them; twist their heads when they hear a new sound; imitate bird calls; indeed, some-times, they lift dirt to their mouths, because they must *taste* the earth itself. Too often as we grow older, we are taught to *not* pay attention to the wonder around us. The advertising industry and mass consumer cul-ture collude to encourage shrinking the scope of our attention.

The natural history knowledge of the average "primitive" human would put to shame the vast majority of moderns, and even—in its breadth, at least—most contemporary scientists. Indigenous peoples throughout the world continue to function with this outward attention to the particulars of the world in which they live of foremost and un-questioned importance.

"Civilization" began to emerge, in both East and West, when agricul-ture took hold, approximately ten thousand years ago. As we settled in to tend crops and raise livestock, our attention tended to narrow to a smaller slice of biodiversity (promoting that which was the most useful while routing out that which was threatening), and social classes be-came established, with fewer people directly involved in the hunting and gathering of food. (For this reason, human ecologist Paul Shepard referred to the onset of agriculture as "ten thousand years of crisis.")

By the time of classical Greece, written language was long established, affecting the way we connected with, and thought about, nature. Aristotle and Theophrastus began to inquire carefully into how animals and plants were put together, and how they functioned. As in so many things, the highlights of Greek culture were transferred to Rome. About two thou-sand years ago (shortly after the death of Jesus), a Roman scholar and philosopher, Pliny the Elder, assembled the first ever encyclopedia, and in the process coined the term "natural history." This monumental work was titled *Historia Naturalis*, typically translated as "natural history." It consisted of thirty-seven books in ten volumes, covering everything from plants to animals, from rocks on earth to stars and planets in the heavens, from art to mineralogy, ethnography to mathematics. In short, it was an attempt to capture all that was known about the world at the time—an ambitious, comprehensive presentation of how the world works.

Substantial confusion exists because of the common translation of *historia* as "history"—leading many to assume that natural history is concerned only with the past (a view amplified by the prominence given to dinosaurs in many museums). But in Latin (and contemporary Spanish), *historia* has two meanings: "history," but also "story." A perusal of Pliny's thirty-seven books makes clear that when he coined this term he was concerned with far more than past events. In many ways, then, a more accurate and telling translation would be "the story of nature."

In the past few years, numerous definitions have been offered for "natural history"—many strictly concerned with biology—but what all have in common is that natural history focuses on what can be seen, heard, or otherwise perceived directly with our senses. It is fundamentally a practice of observation, description, and comparison. My own definition is: "a practice of attentiveness and receptivity to the more-than-human world, guided by honesty and accuracy." Simply put, natural history is the practice of paying attention. This more expansive framing accords with the notion of natural history as story. It's worth emphasizing that, according to this definition, it is a *practice*—a *doing*, not just an inert body of accumulated facts. Natural history is a verb, not a noun.

Losing and Re-Finding the Thread

After the Roman Empire collapsed, the Church became the primary arbiter of what composed valid knowledge in Europe, and its authority discouraged natural history inquiry. The jubilance of outward attentiveness became severely restricted, at times even punished. Thus, natural history in the Western world was hobbled for centuries until the creativity of human curiosity could be repressed no longer, and intellectual explorations outside of Church dogma (and outside, period) reopened in the late Middle Ages and into the Renaissance, when natural history (often, tellingly, equated with natural philosophy) burst out of the gates and began examining and recording all manner of phenomena once again. Natural history began to achieve social acceptance and wider participation in Europe in the seventeenth century. During the spasm of global exploration in the eighteenth and nineteenth centuries, new generations of naturalists avidly pursued the discovery, collection, description, and naming of new plants and animals, bringing stories and specimens home to European museums. The Linnaean revolution in biological taxonomy in the mid-eighteenth century stimulated a boom in descriptive natural history in the nineteenth century. Linnaeus's binomial system provided a simplified and orderly framework for naming new discoveries, and offered a convenient mechanism by which naturalists could claim lasting credit for their work. All the while, empirical realities revealed by naturalists provided irrefutable evidence that contradicted long-held mistaken assumptions about the world we lived in. For example, description of fossils in British hillsides forced us to expand our sense of time. Descriptions of plants and animals, along with tangible specimens placed in new natural history museums, alerted people to the fact that the world was wonderfully more diverse than they had ever dreamed, and that the Middle Eastern fauna and flora portrayed in the Bible represented only a fraction of Earth's living plenty.

––––––

It is critical to pause here and acknowledge that the foregoing describes the arc of natural history in dominant, European-based Western culture, and that much of this development was profoundly harmful to the peoples and places visited by explorers. That the history of natural history in the West was enfolded tightly within an all-too-familiar contemptible saga of colonialism and exploitation. Sophisticated indigenous understanding of home places—what can accurately be described as natural history and ecology—stretched back to antiquity, and has endured continuously to the present. In many ways, contemporary science is only now beginning to catch up to insights that were old before the "new" exploration of the world in the seventeenth and eighteenth centuries. Approaches such as "Traditional Ecological Knowledge" exemplify this.

––––––

Natural history was our species' original form of disciplined scientific inquiry, and it is the ancestral discipline of ecology, geology, paleontology, and cultural anthropology. All these fields bit off pieces of natural history, narrowed their focus, and specialized, creating their own (often exclusive) vocabularies in the process. (It's worth noting that the term "natural history" predates the word "scientist" by eighteen centuries.) Natural history also initially provided the essential context for visual arts, as seen in cave paintings and patterns on pottery, and the oral literature that is storytelling.

Ecology, in particular, was built upon a foundation of natural history. Indeed, the first textbook in the field, Charles Elton's 1927 *Animal Ecology*, began with this clear statement: "Ecology is a new name for a very old subject. It simply means scientific natural history." In the early decades of the twentieth century in the United States, natural history, or "nature study," was seen as an important part of teacher training. However, by 1938, Aldo Leopold was concerned enough about the marginalization of natural history in academia that he delivered an address entitled, "Natural History—The Forgotten Science." He criticized the new wave of science that increasingly took things apart, but failed to explain how they were connected. He bemusedly observed that, should we drop in "on a typical class in a typical zoology department, we [would] find there students memorizing the names of the bumps on the bones of a cat." It is important to study bones, he continued, "But why memorize the bumps?" The sidelining of natural history accelerated after the Second World War, amid the mixed elation and fear of the space race, the excitement regarding molecular biology, and the hubris of the so-called Green Revolution in agriculture. Many ecologists were eager to distance themselves from descriptive natural history and align themselves with the seeming glory of theorizing and mathematical modeling.

Increasingly, the work of academic ecologists was only considered valid if it involved hypothesis testing—"the hypothetico-deductive method," idealized by Karl Popper, who, ironically, was a philosopher, not a research scientist. Hypothesis-driven research has unquestionably yielded essential contributions to our understanding of how living systems work. But valuing *only* hypothesis-driven research has been problematic. Biologist George Bartholomew pointed out that the Popperian ideal of hypothesis testing was modeled on physics and chemistry, and was never as applicable to biological sciences, simply because natural selection is a chance-driven, inherently unpredictable process. As hypothesis-driven theoretical science became rewarded disproportionately in academia (e.g., hiring and tenure decisions) and funding sources (e.g., the National Science Foundation), descriptive science and long-term environmental monitoring became tougher to undertake.

Ironically, given the elevated status of theory in academia, many of the most important theoretical breakthroughs in ecology and evolutionary biology have been based directly on insights gained from field natural history. The two conceivers of the idea of evolution through natural selection, Alfred Russel Wallace and Charles Darwin, were both committed naturalists. Wallace's painstaking observations in Malaysia and Darwin's in the Galápagos were carefully transferred into natural history field journals. Reflecting on their journals led each man to the concept of natural selection, now the cornerstone of biological thought. One of the preeminent theoretical ecologists of the twentieth century, Robert MacArthur, was a dedicated and skillful field naturalist. His groundbreaking study of warblers along the coast of Maine, which radically altered ecologists' view of biological competition, was based on meticulous natural history—hundreds of hours of field observation. E. O. Wilson, arguably the foremost biologist of his era, titled his autobiography *Naturalist* (not "Theoretician"). The bottom line: theory can only be built upon a solid foundation of accurate empirical observations.

One other point: the simple tools needed to do natural history make it among the most egalitarian of sciences. We don't need gas chromatographs or spectrometers to conduct natural history. For this very reason, natural history is not favored by today's academic administrators: grants large enough to build institutions and administrator CVs can't be achieved by funding a few butterfly nets, hand lenses, or pairs of binoculars. That said, contemporary naturalists sometimes do utilize new technologies, such as radiotelemetry, digital photography, and GIS, as they seek to answer the timeless questions of natural history: *Who? When? How many?*

Why Does It Matter?

Natural history is vital in at least three ways: it provides the empirical foundation for natural sciences; it provides the basis for conservation; and, not least, it inspires the most positive aspects of our humanity. This practice of attentiveness encourages our conscious, respectful relationship with the rest of the world, affirms our sense of beauty and wonder, and helps us to see the world, and thus ourselves, more accurately.

Careful observation and description—the cornerstones of natural history—are the basis of all good science. Scientific ecology, to a large extent, dwells in the world of the unseen: abstractions, principles, theories, explanations. There are critical roles for both theory and natural history empiricism, but two points bear emphasis. One, that ecological theory cannot emerge without the direct sensory gathering that is natural history. And two, that natural history is more egalitarian—it doesn't require the book-learnin', and the exclusive language of ecological theory, which typically requires years of formal, sometimes prohibitive, education. Of course, formal training in natural sciences, visual arts, and literature can be a very useful background—but careful observers without academic pedigrees routinely make important contributions to our collective understanding of the world. This truth underlies citizen science programs, such as the Christmas Bird Count.

Conservation has always depended directly on natural history. How can we save species from extinction if we don't know where they are? (Or if we haven't had enough direct experience with them to care?) Indeed, a petition to list a species for endangered species protection in the US requires such natural history details. How can we prioritize biotic communities for protection if we haven't mapped patterns of vegetation? The earliest formal conservation policy in the United States was the regulation of hunting seasons—closures motivated by natural history observations of decreasing game populations. Biodiversity surveys, such as the Natural Heritage programs of The Nature Conservancy, are highly coordinated natural history projects.

Such surveys are essential guides for conservation planning. Conservation is contingent on answers to fundamental questions—*Who lives where? What is the habitat like? How many are there? How does it survive and reproduce?*—the very questions that natural history has been addressing since the earliest humans.

Perhaps most importantly for the future of the world, natural history helps make us better people, by fostering humility and awe, and offers the capacity to build better human societies, ones that are less destructive and dysfunctional. Human psyches engaged with delight, beauty, curiosity, and wonder are much less likely to act out like emotional toddlers. Our current flood of social dysfunctions—violence, depression, anxiety, alienation, and more—would have far less fuel. Natural history engenders humility and open-mindedness, as it humanizes and grounds us by offering larger perspective on the world. This practice makes us better, psychologically healthier people, who are capable of being better citizens. Those who practice natural history are less likely to be myopic and less inclined to believe in the myth of human dominance. Natural history, then, is an important part of how we save ourselves.

Natural history offers a path of kinship and compassion. Psychologists have described the phenomenon of psychic numbing, our human tendency to withdraw our attention from traumatic realities we feel powerless about. We often can't bring ourselves to respond to mass, global problems, but we are much more likely to respond to specifics. Listening to a single person's story—a tale of fleeing a tyrannical government, say, or a loss of a loved one during a pandemic—invites compassion in a way that a news story of genocide, war, a mass shooting, simply cannot. Terry Tempest Williams quoted Mother Teresa: "If I look at the mass, I will never act. If I look at the one, I will." This is equally true when encountering nonhuman others. Crises in "the environment" or "nature" are too abstract for many people to provoke response. But witnessing individual lives—this male kestrel swooping to the top of this Saguaro cactus; this *Phlox* flowering atop this rocky alpine ridge—promotes a sense of kinship grounded in the particulars of these specific lives. Natural history, then, as a practice of attentiveness to the specificities of the world, is a path of compassion—of "feeling with."

We are what we pay attention to. Too often, our attention is deflected in a hall of human mirrors and self-regard. When we choose to pay attention outwardly, we can become intimate with the world we live in, more sincerely befriend our fellow beings, and become more likely to discover our *querencia*—our beloved home ground, "a place in which we know exactly who we are."

The practice of natural history—open-ended exploration of the living world—is the gateway drug to *biophilia,* our innate human capacity for loving the world. As E. O. Wilson, who coined this term, described his own pathway to becoming one of the world's foremost naturalists and evolutionary biologists: "the world of mystery awaiting exploration in ecosystems and the countless species composing them . . . is the opposite of STEM [science/technology/engineering/math education]. It comes from first loving the outdoors, and then taking a scientific approach to natural history." Immersion and falling in love with the world—natural history—first. Later, the rigors of formal scientific training can add depth.

Recently, there has been a rediscovery among some theoretical scientists that natural history is foundational to our understanding of how the world works—that it is truly the empirical bedrock on which all our greatest insights must be built. Moreover, there's a growing recognition that the importance of natural history transcends mere science alone. The new, reintegrated science is infused with arts and humanities— which is to say, science imbued with creativity and deep thought. The rational mind is valued, but not exclusively. Storytelling about nature is too important to leave exclusively to null hypotheses and data tables. Engaging head and heart are both essential.

Entering the Templum

And so we know what natural history is, and why it matters. But how do we do it? Answers are as various as humanity, but a few commonalities emerge when we walk out the door—into *templum,* the open space of observation. In some cases, we have a quarry in mind. We may be hunting for a rare aquatic snake in this canyon emerging from the red cliffs at the edge of the high plateau. We may be hoping to enlarge our list of bird species on this island. Or, we may want to know who is this caterpillar eating the tomatoes in the garden, or seek to identify the woody plant emerging from an urban sidewalk crack.

Oftentimes, we seek with complete openness—with Zen-like beginner's mind, embracing all possibilities of encounter. But sometimes we guide our attention, intentionally or not, by the gear we carry. The simple tools of natural history can function as filters on what we witness. If, for example, we carry a butterfly net, it tends to focus us on these wavering presences wafting on the wind. We net the insect, examine it close up, inspect details of color patterns on forewing and hindwing, shape and color of the antennae, the number of visible legs, and more. Then, we open our palms and release this remarkable life back into the air of freedom. Without the net, we might yet observe butterflies—much can be seen through binoculars—but we are less likely to concentrate on this group of critters. Similarly, which field guides we carry can influence how we perceive the world.

As I have considered my own behaviors in the practice of natural history—and that of hundreds of students, friends, and colleagues over the past several decades—a few fundamental principles have emerged that can provide guidance:

- *Pay attention.*
- *Trust yourself.*
- *Believe that what you witness matters.*
- *Record your observations in notes or images so that they can be shared.*
- *Recognize there's a whole lot we don't know—and that is wonderful.*
- *Humility is a virtue.*
- *Gratitude, above all.*

In the natural history templum, it helps to consciously cultivate *receptivity*—openness to whatever the world may be offering in a given time and place—and *inspiration*—literally, breathing in, fully sensing the place and experience. During, or often after, a field experience, *reflection*—drawing, writing field notes, inquiry into published information, learning from our elders—deepens our understanding and heightens our joy, while allowing us to transmit this understanding to others.

FIGURE 16. Students absorbed in learning how to identify flowering plants, Kluane National Park, Yukon Territory, Canada. © Thomas Lowe Fleischner.

Singing the Old Songs

Sky crystalline blue. The only sound—wind. Scarlet flower clusters bounce at the drooping tips of the long wands of ocotillo stalks. I follow faint trails of coyote across these cholla-studded terraces. Every so often, my head is turned by a lizard darting for cover, the sharp call of an unseen bird, a shift in shadows.

Natural history is our prayer and our celebration, our sacrament and our unkempt affection, our most single-minded examinations and our most wide-open wandering gaze. Sensing the world in four dimensions—zooming in to inner-flower closeness and out to gauging whole mountainsides, casting attention over a whole canyon system, pondering how it came to be.

The practice of natural history takes us, with fistfuls of scree, up into the mountains, smelling the thinning of air, the evaporation of snow-melt. Takes us face-to-face with vipers, or counting stamens inside a bell-shaped flower. Wings lift up off the shimmering mudflat; we witness sandpipers stitch continents together with their long-distance flights. Wading into an equatorial wetland, hands suddenly wrestling the tail of an anaconda. Drifting on the surface of the sea, as orcas surround our boat, reminding us viscerally, we are not the top of this food chain.

Annie Dillard once wrote, "Beauty and grace are performed whether or not we will or sense them. The least we can do is try to be there." That seems to serve as a sort of hymn for practicing natural history. Get out there, be attentive, occasionally be overwhelmed.

Legendary jazz musicians Wayne Shorter and Herbie Hancock wrote an "Open Letter to the Next Generation of Artists" (a message which they clarified was really for everyone). They concluded: "We hope that you live in a state of constant wonder." Natural history—like inspired improvisational music or abstract painting—assures our potential for connecting with wonder—in the familiar and the exotic, in the near and the far.

In this increasingly human-centered age, we maintain our curiosity through an act of will—by turning away, at times, from the merely human—especially from the flat screens of computers and phones which shrink our fields-of-view and collapse all depth into two-dimensionality, while maintaining the grand pretense that human beings represent the complete story. Instead, we can enter the naturalist's trance, which indigenous ecologist Robin Wall Kimmerer has described as "a state of heightened awareness in which we can see and experience the world with extraordinary acuity, all senses engaged and from which an expansive awareness emerges, an experience of connection and meaning making." She goes on to point out that "science, art, and prayer all have this in common, the practice of deep attentiveness, which changes us and then changes the world."

Attentiveness outward, attentiveness inward. Our skin is a very porous membrane. This wild world can heal us, if we let it. Natural history takes us so much deeper into the medicine.

11

Seeing, Feeling, and Knowing Nature

Martha L. Crump

WINTER OF 1971 in Lawrence, Kansas, found me writing my master's thesis while battling mononucleosis and studying for my PhD qualifying exam. But even more angst-provoking, I was seeing my Brazilian study sites—the várzea, igapó, and terra firme forests—not as the enchanting Amazonian paradise I had called home for nine months, but as equations, formulas, and theoretical ecology models. In becoming a scientist, I was neglecting my emotional attachment to nature and the value of natural history. This realization was particularly distressing because since childhood I had felt a close bond with my natural surroundings.

Shortly after passing my qualifying exam, I packed my bags for Costa Rica and the Organization for Tropical Studies field course. I spent delightful hours in the rainforest of the Osa Peninsula, watching little red frogs with blue-green legs (granular poison dart frogs) belt out their advertisement calls to maintain territories and attract females. Best of all, I observed several sequences of courtship and fertilization of eggs that occurred without the normal sequence of events in which the male clasps the female. I had seen a behavior that no one had ever recorded in these frogs! The thrill of discovery felt magical. This intimate experience rekindled my appreciation and passion for natural history—and set the stage for my response to one of this book's driving questions: "How can we keep a curiosity about and love of nature and wild things alive in an increasingly human-defined age?"

Well over a century ago, John Muir wrote: "Climb the mountains and get their good tidings. Nature's peace will flow into you as sunshine flows into trees. The winds will blow their own freshness into you, and the storms their energy, while cares will drop off like autumn leaves." Euphoric effects such as those promised by Muir might be a bit hyperbolic, but numerous studies have revealed that being outdoors boosts physical health, mental well-being, creativity, and productivity. A connection with nature offers both calm and joy.

And yet, fewer people engage in nature-based activities now than a few decades ago, and those who do participate spend less time outdoors. Instead, we are glued to technology. Smart phones, tablets, and laptops keep us in constant contact with friends, allow us to buy practically anything with a click of a button, and provide instant access to world news. Does it matter that people are becoming disconnected from nature?

Yes. It matters because humans are increasingly reshaping nature, and not for the better. Our footprint stretches from the ocean depths to the upper atmosphere. We have destroyed forests; overfished our oceans, lakes, and rivers; reduced biodiversity; effected disastrous climate change; littered and polluted our landscapes, oceans, and waterways; facilitated the spread of invasive species; and recklessly consumed natural resources. For decades, scientists have warned that we must mitigate our impact on the environment—or face systemic failure of ecosystem processes. By now we should understand the consequences of our actions, yet every year our footprint expands wider and deeper, reflecting our individual and corporate greed of valuing immediate material "needs" above the environment. If we are to diminish our footprint and heal the wounds we've caused, we must love nature.

I suggest three approaches as a pragmatic and experiential response to the question of how we can sustain our innate curiosity and love for nature. First, we can rekindle awe and wonder of our natural surroundings by trying to see nature through the eyes of children. Second, we can strengthen our emotional connection with nature through mindfulness techniques and our understanding of nature through close observation. Third, we can advocate for nature and encourage others to become actively involved in respecting and protecting Earth.

> One of the joys of being a grandparent is getting to see the world again through the eyes of a child.
>
> —DAVID SUZUKI, CANADIAN
> ENVIRONMENTALIST AND AUTHOR

Imagine you are watching a caterpillar for the very first time—a multilegged animal with white, black, and yellow-green stripes around its body and two sets of black "antennae." If you are an adult, you might ask: Is this an insect? If so, what kind of insect? But if you are a child, the animal is magical. You might ask: Can I touch it? Will it bite me? Can it see me? Where are its ears? Does it have a nose? What are those wriggly black things sticking out of its head and butt? Is it a boy or a girl? Can I keep it for a pet?

Questions about nature asked by adults versus children can be miles apart, largely based on the difference in experience. Curiosity, the drive to seek new information, to learn about what we do not (yet) know, is an essential human attribute for exploring and making sense of the world. This is especially true for children, for whom everything is new. Children learn about their surroundings through exploring, asking questions, and hands-on experiences.

Walk outdoors with a child and you will be amazed at their engagement with nature. There is no destination. A walk truly is a journey, an exploration. A child will turn over rocks and logs to see if something alive might be hiding underneath; examine puddles for anything that moves; poke sticks into crevices in search of creepy-crawlies. A tree is meant to be climbed; colorful pebbles are meant to be stuffed into pockets; a frog is meant to be caught; and a flower is meant to be picked. Everything is meant to be touched. Sticks become swords and wands, players in a game of imagination and fantasy.

I learned firsthand the rewards of nature exploration with children when, in 1965, I worked as the nature counselor at a camp in the Pocono Mountains for disadvantaged children from New York City. Very few of these inner-city children had ever seen a forest, but they were infinitely curious. During our walks to collect treasures for the Nature Nook—feathers, nuts and seeds, rocks, empty snail shells, and bits of moss and lichen—they asked a never-ending string of questions. When I picked up a toad, knowing it would pee on me, I expected the kids to laugh. They did, but after I explained that toads pee on predators as a defensive behavior, they bombarded me with questions, the most urgent of which was: What if the toad that just peed on you didn't have any pee left to defend itself?

Why do so many of us stop asking questions about nature as we get older? Can we restore our curiosity about our surroundings? I believe we can, by making a conscious effort to view nature as a child does. We tend to think of learning and nurturing in one direction, from adult to child, but we have much to learn from children. Let's allow them to help us reconnect with our curiosity, to help us see the natural world as though for the first time. Spend time with a child outside. Go on nature walks and explore together. Pay attention to what children see, hear, and feel; their questions; and their joy and wonder. Encourage other adults to spend time with children outdoors.

Many years after my camp experience, I had the privilege of sharing my knowledge of nature with my children; they shared their wonder and questions with me. We gathered fossil sharks' teeth and dugong ribs from our creek in Florida; raised tadpoles and released miniature frogs back into the wild in Costa Rica; roamed with penguins and sea lions in Argentina; and recorded seed preference by harvester ants in Arizona. Now, sharing nature with my granddaughter is the ultimate satisfaction in my eighth decade.

Forget not that the earth delights to feel your bare feet and the winds long to play with your hair.

—KHALIL GIBRAN, LEBANESE-AMERICAN
WRITER AND POET

FIGURE 17. As I watched my granddaughter, Fionna McKree, observe and contemplate the first live starfish she'd ever held, I too saw the starfish with fresh eyes. Photo by the author.

When you see your first golden-yellow daffodil of spring, do you burst into a smile? Does watching early morning mist rise from a lake take your breath away? Do you stop and watch the little guys—the fastidious spider repairing its web in the corner of your living room; the otherworldly praying mantis hunting in your garden; the naked slug sliming its way across your neighbor's stone wall? Do you finger the silky, fuzzy nubs of pussy willows? Do you inhale deeply through your nose to savor the intoxicating fragrance of a bed of roses? Ask yourself: Do you feel an emotional attachment to nature?

Although we bond with nature in many different ways, the most profound is through our emotions. The reason varies with the individual. This might be because we are inspired by the diversity, complexity, and transcendent beauty of our natural surroundings. We might feel a spiritual connection with the land, or a profound moral obligation to revere and care for nature. Our natural surroundings might give us a sense of place, of feeling grounded, of belonging to a particular landscape and its flora and fauna. Some of us connect emotionally with nature through outdoor recreation: skiing, fishing, hunting, bike riding, and hiking.

It isn't an understanding of ecosystem processes or concern about climate change that is likely to cause people to bond with nature. It's being out in it—the emotional attachment. And, in fact, studies have shown that time spent outdoors during childhood or adolescence, and parents or other adult role models who fostered their interest in nature, were the two most important factors that have influenced conservationists and environmental educators to embrace their environmental values and philosophies. The underlying influence was not awareness of environmental degradation. As expressed eloquently by David Sobel, "What's important is that children have an opportunity to bond with the natural world, to learn to love it and feel comfortable in it, before being asked to heal its wounds." Adults also need an emotional connection, a feeling of being part of (not separate from) nature, before they can be expected to protect the planet.

We can make a conscious effort to connect emotionally with our surroundings through mindfulness techniques. Here is one to try in your backyard, neighborhood park, forest, or anywhere else outdoors you choose. After settling into a comfortable spot, slowly take five deep breaths while becoming familiar with your surroundings. For the next twenty minutes, use all your senses and focus on your immediate environment. If you see a tree, reflect on its shape, smell, and color. Are the leaves rustling in the breeze, or is the air still? Imagine the anchoring root system; the network of fungi surrounding the roots; and the earthworms and beetles aerating the soil. You might see lizards basking on rocks or bees gathering nectar and pollen. You might hear squirrels chittering overhead; birds announcing their territories with song; or silence. You might feel the breeze on your cheeks and the sun on your neck. Smell the pungent odor of sagebrush, the evergreen scent of pine trees, or the perfume of honeysuckle. When your mind wanders, as it inevitably will, gently refocus on your breathing. There is no expectation of epiphany. This is simply you observing and connecting with your surroundings.

Some people feel a strong desire to know, as well as to feel, nature. For those, the bond with nature is enhanced through natural history.

Literally "the story, description, or investigation of nature," natural history is the attempt to understand and describe nature through observation. It involves the search for patterns and the description of those patterns. Natural history does not require formal training, and in fact many of the most knowledgeable natural historians are self-taught.

Scientists, educators, writers, visual artists, musicians, social scientists, and philosophers have long promoted nature study. In the mid-nineteenth century, Louis Agassiz advised, "Study nature, not books." The Nature Study Movement in the US had its peak between about 1890 and 1920 and was led by such progressives as Anna Botsford Comstock and Liberty Hyde Bailey, both based at Cornell University, the hub of nature study at the time. Its mantra was Agassiz's "Study nature, not books." Bailey wrote that nature study "trains the eye and the mind to see and to comprehend the common things of life; and the result is not directly the acquiring of science but the establishing of a living sympathy with everything that is." One of Rachel Carson's lifelong passions was encouraging young people to study nature. The renowned American architect Frank Lloyd Wright advocated: "Study nature, love nature, stay close to nature. It will never fail you."

Close observation of poison dart frogs in Costa Rica renewed my thirst for natural history. My fondest memories of subsequent fieldwork over the past fifty-plus years are of weeks spent crouched on the ground or perched on a boulder, binoculars glued to my eyes, watching Darwin's frogs, harlequin frogs, and golden toads. In addition to the challenge, and satisfaction, of trying to understand why these frogs do what they do, the hours of observation provided intimate experiences with other animals. An inquisitive neotropical river otter approached to within five meters, stood on its hind legs, and stared intently. An iridescent blue morpho butterfly landed on my shoulder and stayed a while. A shaggy Chilean rose tarantula crawled up my pant leg, groomed itself, and then descended to the ground. Close observation made me truly feel part of nature.

I suspect that most readers of this book already pay close attention to their surroundings. But if you need a nudge, know that anyone with curiosity can become a natural historian. High-tech scientific equipment is not necessary, although a magnifying lens and binoculars can be useful.

Pay attention to phenology—flowers blooming, leaves changing color, frogs calling, geese heading south. Watch birds at your birdfeeder and note which species display dominant behaviors. Go on a "bio blitz," identifying as many flowers, insects, or birds you can spot in a given time period; do this in multiple seasons or different times of the day and compare your results. Sit on a rock or log and observe some aspect of nature from all angles and aspects. Think about point of view. How would an ant view the decomposing log near you? How would an eagle flying overhead view the log? Keep an illustrated field journal and sketch what you observe—it forces you to pay close attention. Express your observations in whatever creative form your talents lead you, from writing haiku or essays to composing music or painting with watercolors. Be a role model and encourage others to become natural historians. Close attention to nature leads to better understanding of our surroundings, but it also offers a perspective that can challenge anthropocentrism.

Pessimists might suggest it folly to think we might ever see a wholesale increase in nature appreciation among the general public, but it is already happening. As I write these words, we are still in the midst of the COVID-19 pandemic that began in 2020. A common feeling people have expressed during the pandemic is the need to get outdoors and connect with nature. There is indeed something magical about being outdoors that eases stress and improves mental well-being, especially in times of crisis, whether fishing in solitude or hiking with friends. I used to see one or two people on my hiking trails pre-pandemic. Now my trails are overrun with people young and old, hiking, biking, and walking dogs. Gardening retail sales have surged, fueled by veterans rediscovering their love of gardening and novices celebrating a new outdoor hobby. In a study conducted by researchers at the University of Vermont, based on online surveys completed by over 3,200 adults, Vermonters reported increased engagement in outdoor activities during the pandemic as compared to the previous year, for example: taking photographs or doing other art in nature (54% increase), gardening (57% increase), relaxing alone outside (58% increase), watching wildlife (64% increase), and walking (70% increase). How quickly we were able to expand our relationship with nature practically overnight! Of course, the question is: Will we continue to seek an intimate relationship with nature once the pandemic is behind us? This seed of nature appreciation that has germinated for many people must be encouraged to grow. We have too much at stake to lose this opportunity.

> Never doubt that a small group of thoughtful, committed citizens can change the world; indeed, it is the only thing that ever has.
>
> —MARGARET MEAD, AMERICAN CULTURAL ANTHROPOLOGIST

By the year 2050, Earth will surely look different than it does in 2024. There will be more of us, demanding more natural resources. We will have lost even more species and wild places. More landscapes and waterways will be degraded. Climate change will have caused millions of people (and many more millions of other animals) to migrate because of rising sea levels, flooding, droughts, and wildfires. The critical question is *how much* change will there be, and *in what ways*? As the primary drivers of environmental change, the answer lies largely in our hands.

We need to change our mindset and our ways *now*—embrace zero population growth, eat more insects and less beef, use more sustainable forms of energy, reduce our consumption of natural resources, and all those other actions scientists have been advising for decades. But the bottom line is that to invest in environmental stewardship, as individuals we must care. How can nature lovers motivate the general public to sustain its bond with nature strengthened during the pandemic? Perhaps the most important (and simplest) action that nature lovers can take is to share their passion with neighbors and other community members. We need to advocate for nature. In doing so, we need to make nature engagement positive, personal, and local—not some abstract concept.

The possibilities are infinite. If your passions and talents lie with visual arts or music, incorporate nature into your creations. If you gravitate toward environmental activism, organize projects and involve the community—plant trees, remove invasive weeds, maintain hiking trails, start up a community garden. If you enjoy working with children, be a mentor. Volunteer to give talks and lead nature walks at your local schools, libraries, museums, and nature centers. Start a neighborhood nature club. If your backyard is the focus of your attention, offer to help your neighbors set up birdfeeders, birdbaths, and bird and bat houses; plant butterfly gardens. Sometimes all it takes for people to overcome inertia is a suggestion and help from a friend or family member. Give talks and participate in outreach events hosted by nature centers, museums, schools, and libraries. If you enjoy writing, write about nature for the general public.

As advocates for nature, I believe we must focus on both the wild and the less-wild. It is easy to love unspoiled alpine meadows and coniferous forests, but what about the backyards, gardens, vacant lots, and urban and suburban parks that most of us interact with more frequently? Can we love less-wild nature? Of course we can. A child's curiosity is just as active while exploring urban nature as undisturbed forest. Adults can do the same. Even in our increasingly human-modified world, there will be the reassuring continuity of progressing seasons; stunning sunsets and double rainbows; mushrooms, flowers, and trees; captivating animal behaviors; lofty rock formations; and spectacular vistas. If we are to heal our wounded environment, we must find beauty and inspiration in the modified nature that surrounds us, in addition to appreciating the magnificence of our national parks and wilderness areas. We need to retain a sense of curiosity and feel wonder about the places we see as "background" to our lives.

But what are the implications of loving human-altered landscapes? First, we need to move past the singular focus of preserving "wilderness." Most landscapes are in fact human-altered to some extent. A 2021 study by Erle Ellis and colleagues reported that people have shaped most of our terrestrial landscapes for at least the past twelve thousand years. Not all human impacts on the environment are negative. In some cases, humans living on the land are actually doing a better job of conserving biodiversity than are the protected and conservation areas that keep people out. Conservation practices of various Indigenous cultures have resulted in fairly intact ecosystems, because their livelihoods depend on living in ecological balance with nature. Such traditional practices include, for example, restricting access to sacred areas, an offseason for hunting and fishing, and leaving cropland fallow to allow for soil nutrient regeneration. Recently, it has been reported that Indigenous-managed lands in Australia, Canada, and Brazil harbor slightly more vertebrate diversity than do protected areas. That finding has profound implications for conservation.

Second, when I suggest that we need to accept human-altered nature, I am not referring to lands ravaged by massive deforestation, strip mining, and the like. I am not suggesting that we bury our heads in the sand and ignore environmental destruction. We need to acknowledge, take ownership of, and then work to heal the environment and restore landscapes where possible and appropriate.

But on another level, does acceptance of less-wild nature threaten to undermine the relentless efforts of conservationists to preserve wilderness areas? For a while now, there has been growing tension between conservationists who advocate protecting nature for its own sake (and from people), often focusing on land preservation and biodiversity, and those who argue for protecting nature for the benefit of people, as in sustainable farming and urban conservation. It doesn't have to be either/or. Many conservationists and natural historians (myself included) believe that there are many ways that people engage with nature, and that if we are to increase public support for conservation we need to focus both on nature for nature's sake and nature for the people. I believe we need to accept and even love human-modified nature if we are to care enough to reverse the damage we have inflicted. At the same time, there is nothing more rewarding and awe-inspiring for me than hiking in the Rocky Mountains of Colorado with moose, bighorn sheep, marmots, hummingbirds, and myriad butterfly species—and never seeing another human being. We can appreciate and respect it all, wild and human altered. In doing so, we will engage with nature more sustainably and lighten our footprint on Earth because we will feel a part of nature, not separate from it.

Postscript

Writers can often point to one individual who provided inspiration. For me that person is the author of this book's afterword, Harry Greene— herpetologist, ecologist, evolutionary biologist, and natural historian extraordinaire. I remember standing in the beer line with Harry at a national herpetology meeting in the late 1990s. He asked me what I was up to, and I told him about the book I was writing—*In Search of the Golden Frog*—about my adventures studying frogs in the tropics. I told him that tales of studying frogs could be just as exciting as those told by macho biologists studying lions and leopards. Harry looked down at me, smiled, and said something like, "You bet!" While inching up closer to the keg, we talked about my ideas for the book and about his experiences working on his newly published book *Snakes: The Evolution of Mystery in Nature*. The glow on his face revealed his passion for natural history writing. After returning home, I reread Harry's book. His intimate reflections on snakes, science, nature, and art allow the reader to see what makes Harry tick—and gave me courage to share my inner thoughts and feelings. Harry showed me by example that personal reflections take natural history to the next level when communicating with the public. Thank you, Harry.

12

Virtual Nature and the Future of the Wild

Susan Clayton

AS WE look toward the future, two discordant trends can be observed in the relationship between humans and the rest of nature. One, stemming from the world of academic research, is a dramatic increase in the evidence that our exposure to nature matters. Being out in nature has benefits for mood, stress levels, creativity, mental health, and social interaction. Even just looking at green or blue spaces can have beneficial effects, not to mention the additional benefits that go along with breathing cleaner air and the opportunity to exercise. People who visit parks more often, or who live in areas with a greater proportion of tree cover, are healthier. Any greenery is good, but areas of greater biodiversity seem to have even stronger effects. This finding is so generally accepted that doctors talk about writing "nature prescriptions."

The second trend is more statistical: human exposure to the diversity of nature is decreasing. Just as we are learning to appreciate the multiple values of nature, we are losing access to it. Biodiversity is plummeting, largely due to the destruction of wildlife habitat but also to pollution, both of which are driven by human population growth, human activity, and human industry. Even if biodiversity were remaining constant, human exposure to it would be on the decline. Because more and more of us live in cities, and work in offices rather than fields, we no longer encounter nature on a routine basis. Because we spend more time watching videos and playing computer games, we no longer rely on nature as much as a source of recreation and entertainment.

Why does this matter? Other essays in this volume talk about the importance of conserving nature and protecting ecosystems. Experiences in the natural world are important in part because of the way they shape attitudes: they tend to promote human interest in, and support for, environmental conservation. I am in complete agreement with this argument, but I want to talk about the importance of personal experience in promoting mental health and psychological well-being. I want to highlight the impacts on people.

What Is Being Lost If People Don't Encounter the Wild?

Many parents, at least in developed nations, worry about their children's restricted access to nature. Richard Louv's 2005 book, *Last Child in the Woods*, prompted a movement, "No Child Left Inside," to emphasize how important it was to get children outside. Without exploring or diminishing the achievements of this movement, I want to focus on the parents' response—their feeling that it was important for children to be out in nature. Although the parents were not necessarily able to articulate their reasons, many of them had an instinctive feeling that their children should have the opportunity to "go wild." Parents were perhaps remembering their own childhood experiences: the joy of squishing in the mud or exploring a tidal pool, the fascination of finding a wild animal like a toad or a turtle and waiting to see what it did. In the past, although these experiences were not available to everyone, they were more widely available than they are today. Money is not always necessary to have access to nature.

Nature provides children with what is in many ways a supportive environment: a context in which stress is reduced, hyperactivity is diminished, and social relationships are more positive. It also provides an important testing ground, where children learn not just about other species but about themselves and their abilities. Can I climb this tree?

Can I dam this creek? Can I move quietly enough to keep this insect from flying away? Can I tell whether or not this rabbit is scared? For children who are lucky enough to have these experiences, they can become part of a persistent sense of self: an environmental identity that situates them within the ecosystem and connects them to its other components. They learn who they are by learning what they can do and who they are like and unlike—children all over the world enjoy pretending to fly like a bird, hop like a bunny (or a kangaroo), or curl up in a pretend bear den. The emotionally rich experiences that they have in nature are also important in providing memories that can be stitched together to form a narrative of identity. They develop a sense of themselves that includes their own relationship to the natural world.

Adults, too, can benefit from experiences in nature to clarify their sense of who they are. In developing and reinforcing an identity, everyone searches for self-understanding. Nature provides the opportunity for a kind of self-reflection that is hard to replicate in other places. People value spending time in nature because it allows them a space apart from the hubbub of the social world to think about their own goals and values. People also want to feel that they are part of something larger, and experiences in nature—especially, perhaps, the transcendent experiences that people sometimes have in the presence of awe-inspiring nature—can provide them with that feeling of belonging and connection. An environmental identity is a self-concept that incorporates a sense of interdependence with the natural world. Like a national identity, or an ethnic identity, it is not the only source of self-understanding: it can be more important in some contexts and totally irrelevant in others. But it is a valuable part of the self-concept.

Research has demonstrated the advantages of an environmental identity. Not only is it associated with a greater ability to enjoy the benefits of nature and with more sustainable behavior, it also seems to be associated with other positive attributes, such as greater well-being and a stronger sense of meaning in one's life. It may even be associated with better social behavior—kindness and altruism—something that researchers have linked to a sense of humility. Some natural scenes present us with a recognition of our own limitations as humans. Simply by confronting a nature that is not under human control, we find evidence of powerful nonhuman forces; by observing vast landscapes, gigantic trees, or uncountable multitudes of insects, we get a sense of our own relatively small size, both as individuals and as a species. This experience of awe and immensity can help us overcome the feeling that we ourselves are the center of the universe, a feeling that may be gratifying, but also comes with a heavy load of responsibility. Paradoxically, realizing our own limitations can be reassuring. And by recognizing our connections to others and to the ecosystem, we are able to act with them in mind and behave in a way that is less self-centered. This is not to imply that nature can magically make us all cooperative and benevolent. Indeed, competition over natural resources is responsible for many serious and prolonged conflicts. But in the absence of an environmental identity, there may be fewer bases for recognizing our connections to others.

With nature increasingly hard to access, what can be done to allow people to develop not only an awareness of nature's value, but also a sense of their own place in it?

Two main strategies are available to us. One is to find ways to get more people outside. Many initiatives are designed with this goal: nature camps, incorporating nature into formal education (e.g., in "forest kindergartens"), and simple parental effort. Visits to national parks are generally increasing in the United States (although COVID-19 disrupted the pattern, 2021 levels showed a large increase over 2020 and suggested the upward trend will resume). These efforts are valuable and will continue to be important. But national parks are not equally accessible or seen as equally welcoming to all groups. In addition, setting aside some areas as "wilderness" while still allowing visitation presents challenges. With more human-nature interaction, the presence of crowds can threaten people's ability to meet their expectations of a visit to nature while also threatening ecosystem health. On the other hand, well-intentioned efforts to protect nature by cordoning it off from human contact may inhibit people's sense that they belong in, and are part of, nature—and relatedly, limit their feeling that it is their job to take care of it. While some protection of natural ecosystems is essential for their ecological health, it does not necessarily promote human health and well-being.

FIGURE 18. A screenshot from "Witcher: Wild Hunt." CD Projekt Red. Reproduced with permission.

Another option is to look for alternatives. People have long coped with a lack of nature by replicating it, or by bringing parts of it into built environments: think of all the fruit, flowers, and vines on wallpaper, carpets, and furniture, not to mention the framed prints and paintings hung on the wall. Some have argued that our affection for pets reflects an attempt to compensate for the day-to-day immersion in nature that was part of our evolutionary history. The popularity of zoos and aquariums also reflects our fascination with nonhuman species. In the current era, we have new tools to mimic nature: nature-based documentaries, video feeds from animal cameras that stream from national parks and zoos; even virtual nature experiences, which will no doubt increase as the technology improves.

It is fascinating, and compelling, to see the extent to which our desire for nature tries to find expression even in the modern world. Even in a video game (World of Warcraft), a survey that asked players about their most and least favorite online setting in the game found that players rated the more natural settings, those containing the most vegetation, as their favorites and showed the least preference for those without any vegetation. Video games are increasingly utilizing realistic and detailed landscapes, accurate enough for players to learn a little bit about the species that are depicted (fig. 18).

Are These Good Enough to Substitute
for Wild Nature?

The ways in which people generate alternatives to nature tend to have certain characteristics. First and foremost, they are managed. Nature videos are directed and produced; zoos and aquariums are designed; video games are developed and created; even animal cams are strategically placed. This means that, rather than encountering nature as an integral part of existence, we tend to make the decision to seek it out, to go (literally or metaphorically) to a place where nature can be found. And the experiences that are sought and shared are not necessarily representative of normal, or typical, nature. People who post nature photos on social media curate those photos and pick the ones that are the most attractive, or the ones most in line with the point their creators are trying to make. In consequence, these images of nature tend to be cuter, prettier, more dramatic, maybe even cleaner than nature itself. So they may establish a baseline expectation that this is what nature should be like.

These alternative experiences of nature are also almost always limited, providing a restricted range of sensory information. Photographs are static and present only visual information; videos present only visual and auditory information. But nature stimulates all our senses; the tactile and olfactory information are an important part of the experience, especially given the close links between smell and memory. The ability of nature to be restorative may be linked in part to this range of stimulation: some researchers have found that technological nature provides some, but not all, of the benefits of real nature. It seems possible that, because we evolved as a species to respond to the natural world and not a simulacrum of it, the chance to integrate sight and sound and smell, and to perceive them all changing in real time in interaction with each other, provides important exercise for our brain, allowing us to develop our cognitive abilities in a way that is simultaneously energizing and (depending on the situation) relaxing.

When we only encounter nature that is pleasing and not overwhelming, there are several possible consequences. First, we may be unprepared for the real thing. Nature in all its messiness seems undesirable and wrong. Even though people say they value wildness, they don't always know what wildness looks like, and in practice people sometimes prefer a more managed landscape to one that is unaltered by human intervention. People may want to remove the parts of nature they find unpleasant, turning wild areas into unthreatening landscapes with no risk of insect bites or unwanted animals. The monoculture of the American front lawn represents the extreme expression of this tendency. These areas are not ecologically healthy, and will not help to protect the ecosystems we all rely on.

Second, we may not be getting the psychological benefits that we would from wild nature. When sociobiologist E. O. Wilson proposed his biophilia hypothesis, he suggested that, because we evolved in interdependence with the natural world, we have an instinctive tendency to relate to it in certain ways. These kinds of relationships with nature may even be important to our mental health (as has been suggested by some ecopsychologists). Certainly there is evidence, as suggested above, that exposure to nature has positive effects on stress reduction, attention restoration, and even social behavior. A plant in a room, a poster on a wall, or a nature documentary seem to have some benefits, but they are unlikely to have the impact that would be obtained from a more immersive environment. Research on psychological restoration suggests that it requires an environment that is perceived as having extent (is not severely limited in either geographic or temporal range) and as being "away" from our regular sphere of activities. Nature experiences that are curated don't allow people to develop the behavioral and cognitive skills that spontaneous experiences may require.

Third, we may be missing out on important ways of developing perspective on ourselves. As described above, nature may provide people with not only self-knowledge but also humility. An environmental identity can provide us with a sense of connection and meaning. If our only encounters with nature are designed experiences for human benefit—a wilderness safari, or rafting the Grand Canyon, or even a nature documentary—the focus remains on the human perspective. We miss the opportunity to feel the sense of our own smallness and limitations.

What Can Be Done?

Many people consider that "wild" means that humans played no part, and "wilderness" must exclude human structures and activity. Bill Mc-Kibben famously wrote about the "end of nature" because there were no longer any areas on Earth that were free from human influence. If we take this point of view, the battle has already been lost. But humans have been affecting nature and altering ecosystems for far longer than was originally recognized. Wilderness, to some extent, is a socially imposed definition rather than an objective way of describing an environment. Rather than taking a dichotomous view—wild nature has no humans, if human influence is present it is not wild—we should consider a more nuanced view and talk about degrees of wildness. This allows us to recognize that people can have a "wilderness" experience while on a guided safari or a whale-watching tour; children can even encounter wilderness while camping out in their own backyards. People can have a shared experience of wilderness.

Rather than protecting wilderness by removing people, we should consider ways to get people *more* exposed to wilderness—but using a generous description of wilderness, one that includes unexamined corners of the city, the garden, or even the home (spiders!). People can discover nature, including "wild nature," in vacant lots, home gardens that allow for biodiversity, or small plots on school grounds. The thing that defines wilderness, from this perspective, is not the absence of humans but the acceptance of the unexpected: there is no attempt to present a particular aspect of nature. Because this can be uncomfortable or even frightening, educators (including parents) should also be deliberate about including an emphasis on skills when introducing nature experiences. Children can learn to be respectful of the wild animals in their backyards, for example, rather than trying to chase them away; they can learn the appropriate ways to coexist with a snake or a skunk or a bumblebee; they can learn to accept the presence of squirrels or deer even when that entails a garden that is less lovely or less productive.

Importantly, experiences of nature mean more than a brief interaction between a person and their environment. These experiences exist within a social context that defines what counts as nature and how it should be valued; they also often include other humans, whether it be fellow students on a school field trip, tourists who have paid thousands of dollars for the experience, or close friends and family members who are sharing an emotionally significant moment. These others help to create as well as interpret the experience. Historically, many people have feared and disliked wild nature and done their best to transform it; even today, one person may see wilderness in what another person recognizes as a managed landscape. Instead of thinking that people's experience of the wild can only be increased by identifying specific areas as wild and then allowing people (regulated) access, we can try to increase experience of the wild by changing understandings of what wild is, and allowing people to find it closer to home.

An authentic relationship with the natural world is built on experiences. These should include not only experiences that are planned and curated to be interesting and pleasurable, but also ones that are spontaneous, interactive, shared with others, and even negative. But we need to be realistic about the ways in which opportunities for nature experiences are changing. There are reduced opportunities for in-person contact with minimally altered ecosystems, but there are increased opportunities for technologically transformed experiences such as documentaries, video games, and, increasingly, virtual reality.

Many people in the developed world perceive themselves as lacking the skills to survive in "real" nature, and perhaps as lacking the motivation to spend time in wilderness. Environmentalists may inadvertently reinforce this perception by disparaging an anthropocentric perspective on nature and insisting the only valid perspective is one that is not swayed by human interests or values. Such a cultural narrative may lead some people to see nature and wilderness as foreign and even hostile, and to define themselves as people who do not like nature. The social context has led them to define their identity as separate from nature. If they can instead interpret their interest in natural scenery, even in a video game, as a valid way of connecting with nature, they may develop an identity in which nature is more prominent. Perhaps a visit to the zoo could be defined as an indicator of an environmental point of view; if it were, there would be more environmentalists in the world. This is not just a semantic shift: people who see themselves as environmentalists are more likely to take future pro-environmental actions. There is also an element of social justice to consider, since some research suggests, at least in the US, that the stereotype of environmentalists tends to depict them as white, with the result that people of color may not see a place for themselves in the environmental movement.

Finding ways for humans to coexist with nature in a way that is healthy for both sides is always a balancing act. We can't hope that people—all people, including urban residents and people with financial constraints—will all have opportunities to enjoy wilderness experiences, especially as the wilderness is vanishing. We can't expect that people will abandon their technology in order to spend more time outside. But we can look for ways to integrate technology with nature that incorporate some of the advantages of real natural experience, and we can facilitate accessible experiences with nature that allow people to develop a sense of themselves in relation to the natural world.

The most practical way to do this may be by using channels that reach the greatest number of people: formal and informal learning environments. Schools in the future will increasingly need to teach students how to use and interact with technology (although any attempt I make here to describe these technologies will rapidly be outdated). Some of these lessons could incorporate the natural world, as a focal point or simply an incidental part of the lesson: science classes, for example, might incorporate the remote collection of data about natural environments; classes in coding could consider challenges in VR representations of nature. Similarly, informal learning programs could use video games as a stepping-off point for nature experiences, and even encourage people to utilize something about those nature experiences in a return to the game. Hopefully, through the creative use of techniques that allow technology and nature to coexist, we can allow the wild to remain a part of human experience.

13

The Digital Animal

Bill Adams

IT IS 7:00 p.m. in the African bush. The sun has set, and darkness is almost complete, closing over the long grass and the acacia trees. The sky glows in the west, turning a luminous blue-black as the light fades. A bull elephant approaches a multistrand wire fence, and deliberately, almost delicately, pulls down the sagging post with its trunk, and steps through.

Behind him lie the trees and scrub of the Rumuruti Forest Reserve. In front lie fields belonging to small farmers, where maize is starting to ripen. The elephant moves forward, hungry. All night he picks his way between the fields, barging through brushwood fences, avoiding the scattered mud and tin houses, eating and moving on. Just after 4:00 a.m., he is back in the Forest Reserve, crossing the fence without effort, to spend the day resting among the trees. The next night, punctually at dusk, he goes off crop raiding again.

This elephant has a name, Genghis Khan, called after the legendary Mongol emperor by elephant researchers working for the charity Space for Giants. They have a program to address the running conflict between small farmers and elephant crop raiders in Kenya. As part of this, they have been trying to understand how and where elephants moved across the landscape of Laikipia, an extensive area of farmland and bush northwest of Mount Kenya.

Genghis Khan started his crop-raiding expedition on January 13, 2011. But I am watching him in August 2021—or rather, I am watching his track unfold as a video overlay on Google Maps on my laptop, and imagining the scene, as he moves from the land where his presence is tolerated, or even welcomed—on the large ranches managed for wildlife—onto the land where he is unwelcome and feared, a scourge of ripening maize and other crops.

I am able to know where and when Genghis Khan went because around his neck he carries a GPS collar. This is a metal box, hanging below his neck on a broad reinforced canvas collar. Fitting it required a low-flying plane (to find him), a four-wheel drive truck, a vet with a tranquilizer gun, and a team of people to attach the collar and look after him while he was unconscious. Once installed, the collar sends a signal every hour through the local phone mast network, giving Genghis Khan's exact position. These data points are transmitted across the internet and end up in a GIS program on one of the Space for Giants computers. There they are analyzed, revealing where the elephant goes and when. Animated, the locations mark his track, into and out of the farmland, night after night.

The problem of crop raiding (and other forms of "human-elephant conflict") is serious on Laikipia and also elsewhere in Kenya, and indeed throughout Africa and Asia, wherever elephants share landscapes with farmers. In recent years, digital data from radio collars has begun to provide conservation scientists with novel insights on the interlocking—and mutually risky—activities of elephants and farmers the day and long night round.

Watching this video, the question that suggests itself is this: How significant is the collar around the elephant's neck and the digital device it carries? It is clearly a new way of observing the elephant, another source of data for scientific study of its life around human society. But is it more? Does the digital device change our understanding of the elephant, and how we perceive and value it? To pose the question more broadly, does digitalization change the way we know and value nature? The answer to these questions is complicated.

———

The immediate observation is that this astonishing technology already seems normal. Indeed, a GPS collar on an African elephant is just one example of a bewildering variety of digital tags and devices now available to attach to living animals in order to yield a stream of location data.

Radio tracking of wild animals dates back to the 1950s, enabled by the invention of transistors at the end of the Second World War. Scientists discovered that transmitters with a unique radio signal could be located by triangulation between fixed receiver masts or (eventually) mobile aerials. They proceeded to attach transmitters to mammals and birds using collars or harnesses, and developed systems of suckers or darts into the body to track fish and marine mammals. For the first time it proved possible to follow animals as they moved through their wild habitat beyond visual contact.

Over the decades, telemetry devices have grown steadily smaller and more powerful, with much longer battery life. Detection is now possible across longer distances, no longer only by a field worker carrying a cumbersome aerial, but from aircraft, helicopters, and even satellites. Moreover, locations are identified not by repeated observations but by GPS, drawing data from overhead satellites.

Technological advance (particularly in battery life and weight) means that it has become possible to track smaller and smaller animals. Miniaturized satellite tags can be fitted to birds weighing as little as 10 g, and scientists report the tracks of migrating arctic terns or foraging albatrosses the length of the globe and deep into the Southern Ocean, where direct human observation would be impossible. Even very small birds can now be tracked through their global migrations using tiny digital devices called geolocators, which can weigh as little as 0.3 g. These consist of just a battery, a light sensor, a clock, and a recording chip, and are stuck to the feathers of the bird's rump. Data on daylength and time are recorded, allowing calculation of the location of a migrating bird every twenty-four hours. To get the data, the bird has to be captured and the geolocator removed. Despite this constraint, scientists have been able to follow the paths taken by birds such as the hoopoe or the nightingale as they migrate south from their breeding grounds in Europe to wintering grounds in sub-Saharan Africa.

Smaller still, lightweight radio tags ("nanotags"), emitting radio signals that are detected by collaborating receiving stations, are used to track animal movements in the Motus Wildlife Tracking System (Motus), an international program led by Birds Canada. The method offers fine temporal precision (potentially a fix every few seconds), allowing detailed analysis of behavior, patterns, direction, and speed of movement. Motus tags have been fitted to small birds, bats, and even large insects. An ever-widening range of species are now regularly tagged and tracked, by land, sea, and sky.

And yet tags are not the only kinds of digital device used to generate location data for animals. One of the most significant is the fixed digital trail camera, or "camera trap" (fig. 19). As costs have fallen, these devices have become increasingly dominant in surveys of the presence and abundance of terrestrial mammals, particularly of rare species in difficult-to-access environments such as forests. Camera traps have moved from supplementing to replacing conventional field surveys of mammals and, in open habitats, airborne censuses. A recent special issue of the conservation journal *Oryx* carried papers reporting the use of camera traps in studies of snow leopards, Indochinese and Sumatran tigers, clouded leopards, and brown hyenas, among others. Camera traps are also increasingly being adopted for nonscientific use, both by sport hunters and amateur naturalists, to establish what animals move where on wildlife trails.

Digital technologies have also revolutionized aerial remote sensing. Since the 1970s, multispectral digital imaging from satellite sensors has provided civilian environmental scientists with relatively affordable and repeated images of the earth within and beyond the human visual spectrum. While many animals are too small to show up on satellite images, increasing spatial spectral resolution allows aggregations to be detected. In 2020, scientists at the British Antarctic Survey reported the discovery of eight new emperor penguin colonies in Antarctica, using data from the European Space Agency Sentinel-2 satellites to identify guano-stained ice. Drones, or UAVs (unmanned aerial vehicles), have brought the cost of such surveillance down dramatically, broadening its accessibility, and allowing low-altitude identification of smaller species. Thermal infrared

FIGURE 19. A trail camera mounted on a solar-powered elephant fence in Kenya. Trail cameras work automatically, triggered by motion, and provide time-stamped digital image and GPS position that locates an animal in a particular place and time. They allow detection of scarce or cryptic species and identification of individuals with distinctive markings (e.g., patterns on big cats, or distinctive ear tatters on elephants). This particular camera trap is mounted on a gate through the twelve-kilometer-long West Laikipia Fence, Kenya, built to stop elephants moving from spaces where their presence is tolerated (on the left of the picture) and onto smallholder farms, where they raid crops. The gate allows livestock movement from one side of the fence to the other. The camera was installed by the Kenyan NGO Space for Giants to see which elephants got through the gate and when— and also how the gate sometimes comes to be left open. Photo by the author.

imaging from low-flying drones can detect warm-blooded animals such as deer against cooler backgrounds at night.

Digital devices are also increasingly used to detect, and therefore locate, animals by sound. Like digital cameras that can operate beyond the range of human vision using infrared light, audio sensors can operate beyond the range of human hearing. There are, for example, mobile phone apps that will record and analyze, on smartphones or tablets via an ultrasonic microphone, the calls made by bats. A system of acoustic sensors has been used in Sri Lanka to detect and locate Asian elephants around villages from their low-frequency vocalizations. Sonic surveillance can be conducted underwater, with whales detected like submarines, as they move unseen through the deep.

———

It is one thing to note the rise of digital devices, and the location data they provide. It is another to assess their significance. What do they *do*? First, and most obviously, they underpin the creation of new scientific knowledge. They provide unprecedented and sometimes intimate detail on individual animal lives. They make the scientific gaze spatial, turning unbounded ecosystems into laboratories. They carry the potential to transform studies of animal behavior and animal ecology, providing data on what animals do when they are not being watched, when they are far away from a human watcher, alone, or in company of their own choosing. They allow the mapping of species distributions in space and time, and of the factors affecting survival and population change. They also enable studies of nonhuman animal responses to humans, and of individual animals in flocks and shoals. They make such studies possible in environments that humans cannot reach, in the deep ocean or in the skies.

The second big thing that the new knowledge created by digital tracking and location does is to make possible better designed and targeted conservation measures. Knowing where animals go enables conservationists to devise strategies to protect them. Satellite tracking of the migratory paths taken by common cuckoos, a species declining rapidly in the UK, showed that they used two different routes to their wintering grounds in central Africa. Those from the most rapidly declining UK populations mostly traveled via the Iberian peninsula and the West African Coast (rather than via Italy and the Sahara). Spatial data start to reveal the hazards of previously unknown migratory paths. The conservation gaze is lifted beyond familiar problems of the breeding area (such as habitat loss or the use of agricultural pesticides) to think in terms of other landscapes, other forms of land use, in other countries' "staging" or wintering areas. The digital data changes the way we think about the cuckoo and its conservation. Its global, commuting lifestyle is revealed: as Mediterranean or African as it is European.

The use of digital devices to locate animals can also affect conservation more directly. Tagged animals can even be recruited as agents in their own conservation. The loss of signal from tags attached to endangered birds of prey has provided evidence of illegal killing on grouse moors managed for recreational game shooting. Albatrosses in the southern Indian Ocean have been fitted with solar-powered loggers that detect radar transmissions. These have been used to identify ships operating with their Automatic Identification System (giving their location, name, nationality, call sign, speed, heading, and activity) switched off. Potentially, tagged seabirds could be recruited as monitors of illegal fishing in the remote ocean: albatrosses need no salaries and no ports or airfields to refuel.

The attachment of digital devices to animals can also have negative impacts on the lives of individuals. Tagged animals become identified as individuals, distinguished from the crowd. The damage done by a collared elephant raiding a farm at night is identified to that individual, which becomes known and labeled as a "problem animal." Its actions are deemed to become the responsibility of the organization that fitted the collar, or the government wildlife authority that granted the permit. A management response is demanded—it might be improved fencing or a compensation program for farmers, but it might also be the shooting of the tagged animal. In 2015 a great white shark carrying an acoustic tag was killed as an "imminent threat" to human bathers, when its track showed it approaching a popular beach in Western Australia. Without the tag, perhaps nobody would have known it was there. The slow-burn benefit to a species of better data on individuals' movements can be trumped by demands for lethal control.

———

And so, to answer the question posed earlier about the significance of digital devices and the streams of locational data that they generate is that they both change things for the animal, and in our understanding of it and response to it. It could be said that digital technologies create digital animals.

The digital animal is linked to the real flesh-and-blood animal, but remains distinct from it. It is not alive, but it moves in actual space, as the digital data that constitute it are transmitted onto the hard disks of scientists' computers and are shared across the earth through the internet and stored in standardized format in archives. The data is machined by analytical software, and reconstituted in scientific accounts of animal movement, published in scientific dissertations and journal papers, distributed to libraries and desks, and disseminated through social media.

The digital animal—large, mobile, and individually identifiable— acquires a charisma through its "trackability." Conservation publicists and fundraisers build their messages around charismatic species, because they make sense to policymakers and stakeholders, and tend to evoke a warm and generous response in potential supporters. Location data are easily collated, presented in the form of videos on YouTube, or regular blog updates. Digital animals attract attention; they draw human trackers in, and turn them into fans.

The critical feature of the digital animal is that it can be identified as an individual, around which stories can be woven. The movements of tagged animals offer powerful imaginative opportunities for observers, to be absorbed into the embodied life of the digital animal on the screen. The movements of individual animals are the source of popular headlines such as "Longest Recorded Non-stop Migration" (a male bar-tailed godwit was reported as having set a new record of 7,500 miles over the Pacific in 2020). They also allow the development of a curated stream of posts about individual birds. So, for example, in the UK the conservation charity the Royal Society for the Protection of Birds (RSPB) ran a program fitting satellite tracking tags to young ospreys in Scotland, to study their movements as they migrated to wintering grounds in West Africa. The birds were named, and their journeys could be tracked on their website, and overlain on Google Earth. The British Trust for Ornithology went one further with their cuckoo-tracking project (described above). Their cuckoos, all named (all male, because the females are too small to safely carry the available tag), were available for sponsorship. The sponsor received online updates from their chosen bird, including when it arrived in Africa in the late summer and returned to the UK in the spring. Sponsors, which are small-scale, have included schools, individuals, and companies.

———

So what is this novel beast, brought into being by digital technologies? To see it, let us return to the elephant. The digital elephant on my laptop screen is an avatar, something representing the actual physical elephant. It moves in the digital world as the flesh-and-blood elephant moves in the real world. The GIS system drapes a digital landscape around it—farms, roads, forest reserves—to match the real land uses and land users that the real elephant encounters.

Yet the digital elephant is more than a representative, slavishly following its real-life self. The digital elephant is irrevocably connected to the real one, but one could say it has a life of its own. It goes places that the real elephant does not and cannot (down those fiber-optic cables, across your computer screen). It appears and disappears when summoned, in times and places far removed from the physical original. Potentially, these movements could be repeated endlessly. Its digital life could be eternal, limited not by genetics, physiology, ecology, and chance, but by the industrial supply of power to server farms and the fickle capacity of aging software. Certainly its existence is completely unmoored from the life chances and lifespan of any actual elephant roaming the farmland and forests of Laikipia.

The digital elephant also has a cultural significance quite divorced from that of the real elephant. Of course, its significance derives from the elephant it represents, the actual animal whose hot and destructive body bore the digital device that generated the flow of GPS points that compose the parameters of the digital elephant's existence. Indeed, it is only the cultural values attached to the actual elephant that makes its digital form significant—its magnificence, its vulnerability, its destructiveness.

But the digital elephant also has significance and latent power of its own. It is the digital elephant whose movements are analyzed in scientific papers, portrayed on YouTube videos, presented to policy conferences about human-elephant conflict. The digital elephant stands in for the real elephant, bearing the burden of stories and debates, rhetoric and policy formulation. The digital elephant strides the conference hall and seminar room, the optic cable byways of the internet, and the human imagination. The real elephant, perilously, walks the bush.

———

Digital technologies give us new ways of seeing, and new opportunities to look into the lives of wild animals for new audiences. But how do they change the ways we think about and understand nature, or the ways we value it?

First, digital technologies make animals knowable in novel ways. We can see where they go when they are "out of sight," beyond the spectral range of the human eye, or identified through sound alone. The digital animal is easier to locate, research, and understand than its real-life alter ego. And what it teaches us helps us to understand the ecologies of actual wild animals. The way in which a digital animal, documented and evoked by technology, can be known at a distance, followed, researched makes it possible to value it—and therefore perhaps love it—in new ways.

Second, the digital animal is always an individual. Unlike aggregate scientific data—about animal populations, or ecosystems—each digital animal is an individual. Its movements, its preferences for one place over another, its experiences—its death—is individual. As such, the digital animal is always the central character in a story, and such stories are compelling. Like *White Fang* or *Bambi*, the individual animal's life, and its interactions with humans, comes ready-packaged as a morality tale. Through the individual, it is possible to understand the wider interactions of human and nonhuman. Potentially, digital animals evoke empathy in ways that actual animals rarely can, and teach broad lessons about human impacts and responsibilities. There may be "heart" in the digital wild.

Third, digital animals are observable by new and diverse audiences. No longer can animal behavior and movement only be seen by someone trained in field craft, or curated by television wildlife presenters on film captured at great cost in far-flung places. Data can be collected, collated, crunched, and presented through digital media to anyone who can log on to a website. Not only does this open up the natural world to new audiences (tech-savvy young people experiencing the world through the screens of mobile phones, for example), but it allows such lives to be broadcast to potentially huge and distant global audiences. People in one country can log on to data streams generated by animals on the other side of the earth—a polar bear's path watched from Nairobi, or a migrating cuckoo from Beijing. This also seems a gain—the digital animal can move far and wide to inform and entertain mass global audiences, potentially creating and strengthening empathy for nonhuman nature.

So far so good: the digital animal seems like a valuable and true extension of the real, capable of informing and sharpening human awareness of the nonhuman, and encouraging the creation of empathy and connection with nature. But this is not the end of the powers of the digital animal.

There is a risk that the digital avatar will supplant the actual animal. To conjure an elephant's path through the African bush by the flick of a finger on a smartphone screen gives instant access to that digital animal. The technology offers an always-available version of the wild, endlessly updated and prepackaged for the social media age. In this form, nature can be compulsively attractive, but is detached from the situation on the ground. Even if Richard Louv's titular "last child" walks in the woods, they will be carrying their mobile phone, and may judge the wild against the alluring familiarity of its digital version.

So, if we learn to see living animals through their digital avatars, there is a potential danger that we will become detached from a sense of the changes in the actual world itself. The very appeal of the digital animal could reduce or even replace interest in and concern for its living wild model. As Colin Ellard suggests in *Places of the Heart*, perhaps there might come to be "nothing very special about the thing itself because at a time of one's choosing, one can dial up an authentic 3D immersive experience that will have exactly the same effects on the senses." We humans might become satisfied with digital versions of the wild, unaware that their real counterparts are gone.

The compelling availability of the digital animal is compounded by its unnatural longevity. The digital animal has an electronic life that is detached from the actual and can endure beyond the limits of a fleshly life. It can live on long after the real animal from which it draws its existence is dead. It can be conjured on screen again and again as if alive, recreated from stored data. Its existence is therefore not closed within the period of organic life, but is available for deep storage in a data vault, to be recreated at the flick of a switch, awakened by the caress of software mining the computer banks of some cloud storage facility. The digital animal can be conjured back to life even if its living model is long gone.

The use of digital technologies therefore offers unprecedented opportunities to bring the wild into the digital world, and use the power and reach of those technologies to extend human understanding of and empathy for the wild. Yet it also poses the risk that its very strengths will lead it to supplant the endless untidiness and visceral physicality of the living world, and offer instead a sanitized archive recording of living diversity that is being driven extinct. There is both potential and peril in creating digital animals by tracking the wild.

14

Hope for the Wild in Afrofuturism

Christopher J. Schell

BIOLOGISTS AND ENVIRONMENTALISTS dedicated to studying the natural world may say that hope is a scarce resource of late. Human-driven shifts to the climate have increased the frequency and severity of extreme climatic events, including droughts, heat waves, flooding, and massive wildfires. Despite a growing recognition and awareness that humans are the primary contributors to the climate crisis, the slow and often tepid political response to counteract such extreme disasters is underwhelming. As once-in-a-lifetime natural catastrophic events become more commonplace, we are becoming increasingly aware of how fragile ecosystem health is and how interdependent we are on the ecosystem services provided therein. Societies globally have a vested interest in functional ecosystems: provisioning services provide food, water, and other natural resources; regulating services balance temperature, precipitation, and other abiotic (e.g., water, air) factors; supporting services maintain nutrient cycling and sustain habitats; and cultural services enrich the educational, spiritual, and recreational values we gain from our natural environments. Notably, ecosystems do not exist solely for human benefit—other nonhuman species also have a claim to a healthy and functioning ecosystem. Conserving ecosystem services—and by extension the natural capital that supports those services—is therefore both a global imperative and a just obligation.

Biodiversity is a central pillar to that natural capital, and unfortunately, human-driven alterations to natural landscapes are contributing to global biodiversity loss en masse. Biodiversity often serves as the conceptual and literal shield for ecosystem health and function: more intact links or *nodes* (i.e., species) within biological communities strengthen the integrity of any biological network. The removal of species deletes nodes within the network, and that makes it less robust and more porous, reducing species interactions that sustain the system. For example, predators that are reliant on an extinct prey species may similarly go extinct. Likewise, the extinction of bees can lead to reductions in pollination services that hinder primary productivity of native plants. Hence, accumulating extinctions can precipitate trophic cascades that snowball to the collapse of biological communities, and with it, ecosystem health. This is especially troubling in the Anthropocene, where climatic and environmental changes can accelerate biodiversity loss, compromising environmental and societal resilience to change.

Society and the natural world are thus at a pivotal crossroads— continue the current status quo of exploitation and resource depletion or develop new frameworks that radically reimagine our relationship to the natural world. Make no mistake, it is an urgent priority that we collectively utilize and implement strategies that will directly address these interconnected calamities while simultaneously building resilience for a challenging environmental future. The damning recent Intergovernmental Panel on Climate Change (IPCC) Sixth Assessment Report illustrates the need for alacrity, as we are projected to blow past the key warming ceiling of 1.5 degrees Celsius with greenhouse gas emissions reaching their highest levels in decades. In parallel, the Anthropocene epoch marks an unsettling start to the sixth mass extinction event on Earth, with an extraordinary loss of biodiversity not seen since the end of the age of dinosaurs. Technological and scientific advances in climate mitigation and biological conservation have propelled some measures to abate climate and environmental degradation, yet do not feel commensurate with the speed of climate-driven shifts. Moreover, the overwhelming feeling of fighting an unwinnable battle against the myriad societal, political, and economic forces in opposition of sustainable practices can exacerbate the fatigue and ecological grief that underpins detachment from producing solutions to the concurrent crises.

Rescuing ourselves from this negative social-ecological tailspin rests in the lessons of the African diaspora, and a fight that transcends space and time. Black communities globally have found hope, love, creativity, and strength in community. Such familial bonds exist across timelines with ancestors and living relatives as part of the larger tribe. The collective struggle to survive, find joy, and fight for justice despite overwhelmingly brutal opposition is an exemplar of resilience. Indeed, from slavery to genocide, Jim Crow segregation to apartheid, the intergenerational struggle for civil rights and equality provides a compelling framework to begin unpacking how we regain our bonds with nature.

Take for example the expression of Black joy, an expression emanating from pride in the identity and heritage of our African roots and one of the most beautiful forms of resilience. Events like the annual Black Joy Parade in downtown Oakland, California, showcase the emotive, vibrant, and diverse manifestations of such joy. The complex tapestry of experiences from the African diaspora are on full display—from the quad-skaters to the steppers, the drumline to the Black cowboys, educational empowerment to voting rights activists. The outfits, the art, and the hairstyles showcase our creativity, innovativeness, and beauty. The beats and rhythms provide a lens into the past and the future, while also celebrating the present. Dancing and smiles by participants and spectators alike signal an overdue release—from stress, grief, and the mental taxation of societal conformity, if even momentarily. Indeed, after a COVID-induced hiatus, the fifth annual Black Joy Parade returned stronger than ever, proudly taking center stage as the largest parade of its kind in California, and arguably in the United States.

Such hope and joy persist despite the concomitant ecological and societal calamities shaping reality outside the boundaries of 17th Street and Broadway. In the last few years, the COVID-19 pandemic has disproportionately claimed the lives of Black and Latinx Americans. Concurrently, unjust murders, disproportionate incarceration rates, and state-sanctioned violence decreed on Black bodies has contributed to the chronic and intergenerational trauma experienced in Black communities. Further still, unjust zoning laws, gentrification, and displacement persistently uproot Black communities, while inequitable exposures to environmental and climate disamenities (e.g., poorer air and water quality, reduced green space) jeopardize health outcomes. Continued disinvestment in infrastructure, unequal access to natural capital, and the lack of economic opportunities widen the racial wealth gap, making it exceedingly difficult to survive as a Black person in America. Such patterns of disinvestment and devaluation are rampant on the global stage as well, as disease outbreaks, geopolitical conflicts, and climate-induced disasters receive a fraction of the attention that similar instances do in countries with less melanin.

The mere existence of Black joy is an act of defiance, one that customarily withstands the onslaught of destruction, disinvestment, dispossession, and death. In this way, Black joy serves as an elixir to rejuvenate communities and motivate the fight for the civil rights of all peoples. This radical inclusivity serves as the foundation of social and environmental justice movements, which stress the humanity of all peoples and the need for equitable access to healthy environments. Consequently, its infectious nature helps to envision what could be, serving as a catalyst for hope despite a grim and uncertain present. Key to understanding the persistence of Black joy against insurmountable odds is the fundamental belief that the spiritual energies of both our ancestors and descendants flows through us. This is typified in the widely known and recited statement in Black communities—"I am (We are) our ancestors' wildest dreams." The phrase originated from New Orleans visual artist, activist, and filmmaker Brandan Odums, and was popularized by influential Black figures like Ava Duvernay, who used the phrase in tribute to the ancestors of First Lady Michelle Obama. Melvinia Shields, who was born a slave in 1844, would be survived by five generations of progeny, ultimately leading to her great-great-great granddaughter—Michelle Obama—who would lead the very nation that sentenced Melvinia to life in bondage. In this example Duvernay and Obama honor our ancestors' fight for survival as stolen people on stolen lands, while constantly working to improve current conditions to fully liberate future generations from systemic oppression. Black joy alone, however, is not an expansive enough discourse to fully articulate the rich complexity found within this example. Unpacking this complexity and its lessons for conserving biodiversity, and by extension our natural world, requires an all-encompassing framework.

Enter the guiding narrative, cultural aesthetic, and philosophical movement known as *Afrofuturism*, which reimagines and explores a world where people of the African diaspora are fully liberated from the physical and cultural vestiges of colonial destruction and white supremacy. This philosophical freedom allows us to explore the intersections of Black culture, science, and technology to consider future timelines and realities that authentically place Black people at the center of global societies and advancement, rather than footnotes to global progress. Though the term was originally coined by cultural critic Mark Dery in his 1993 essay "Black to the Future," artists, writers, and musicians were creating the emerging genre decades before it was codified. For instance, prolific writers such as Octavia Butler (*Parable of the Sower*, 1993) and Ralph Ellison (*Invisible Man*, 1989) have penned multiple Great American Novels that served as biting critiques of classist, racist, and sexist structures, while also highlighting the need to spotlight Black characters as protagonists rather than stereotyped caricatures. Prominent artists from Jean-Michel Basquiat to Kaylan F. Michael composed canvasses that collided African and African American aesthetics with various science-fiction motifs. Further, musicians from Parliament-Funkadelic to Herbie Hancock, and Outkast to Janelle Monáe brilliantly composed albums that blended rhythm and blues, hip-hop, jazz, and techno to birth new genres of psychedelic soul and funk that now are mainstays in American—and global—music culture.

Perhaps the most recent and popular example of Afrofuturism is Marvel's *Black Panther*, a comic book turned movie franchise that envisions an entire nation—Wakanda—that exists as the most technologically advanced civilization on the planet. The character of Black Panther first appeared in the Fantastic Four #52 comic in July 1966, during the founding and rise of the Black Panther Party, a political organization headed by Bobby Seale and Huey P. Newton in Oakland, California, from 1966 to 1982 that stressed black liberation, sovereignty, and independence from white tyranny and brutality. At the time, Black Panther was the first and only Black Marvel superhero that had superpowers, and not surprisingly was lauded as a crowd favorite among Black Americans. To fully understand the gravity of this event, the Black Panther character arose in the heart of the civil rights era, shortly before Dr. Martin Luther King Jr. was assassinated and the Civil Rights Act of 1968 was signed. Fast forward to the late 2010s, and new prominent American novelists Ta-Nehisi Coates and Roxane Gay publish new graphic novel installments of Black Panther that further explore both the technological and scientific marvels of the Intergalactic Empire of Wakanda. Moreover, the 2018 Black Panther film reigns as one of the highest-grossing films ever created by the Marvel Cinematic Universe (MCU) and has sent reverberating shockwaves of pride and empowerment throughout Black communities globally. This is especially evident in Oakland, California, which has affectionately been given the nickname "Oakanda" after the release of the 2018 film.

So, then, how exactly is the Afrofuturism discourse applicable to narratives around the wild and nature? The examples highlighting Afrofuturism above certainly showcase extraordinary creativity and ingenuity, yes, but some may question how that is relevant to an increasingly dangerous collapse of the natural world. If our goal was to amplify connection among people and nature, including its wilder varieties, it would seem counterproductive to envision a more technologically advanced future as that bridge. In essence, how does Afrofuturism help to generate hope and solutions in our current reality? Revealing the relevant core lessons from Afrofuturism to the natural world requires that we transcend the pages, canvasses, musical arrangements, and film studios of the genre and interrogate the figures who created these reimagined worlds. Simply put: the works alone do not reveal the connection; Black people and our embodied realities *are* the connection. The effective coping and adaptive strategies used by Black communities to thrive despite centuries of violence and oppression are the key to building resilience in the natural world. The persistent fight for truth, liberation, justice, even in uncomfortable and challenging contexts, embodies the beautiful struggle—and represents the definitive thread between Afrofuturism praxis and conservation of the natural world.

Afrofuturism at its core is a metaphysical and symbolic struggle to dream of beauty regardless of the current reality. Black Americans have fought to hold the United States accountable to the ideals it espoused in the Declaration of Independence in 1776. As the recent *1619 Project: A New Origin Story* written by Nikole Hannah-Jones illustrates, Black Americans have fought for a country that has repeatedly categorized them as inferior or less than human. As an insult to injury, American society has manipulated historical texts and records to eradicate historical narratives that provide the unabridged legacy of slavery in an alleged free society. Despite revisionist histories and the omission of key details, Black Americans have continued to fight for the civil rights of all peoples. Case in point: activists like the US House of Representatives John Lewis and author James Baldwin have consistently fought for the liberation of Black peoples, though they passed away before seeing the full gravity of their life's work. Lewis, for instance, addressed the need to steadily work toward freedom, stating that "Freedom is not a state; it is an act. . . . Freedom is the continuous action we all must take, and each generation must do its part to create an even more fair, more just society." Baldwin, in his 1962 essay for the *New York Times*, writes that "not everything that is faced can be changed; but nothing can be changed until it is faced." Both Baldwin and Lewis dedicated their lives to activism, which would have required some hope for a better future.

The parallels between confronting systemic racism and the climate crisis are revealing. Effectively dealing with the climate crisis and biodiversity loss requires that we acknowledge our role in driving both processes. Similarly, confronting and effectively eradicating systemic racism means reconciling and healing the past and current transgressions of our American ancestors. In both instances, the reticence to deal with the driving forces of white supremacy and capitalism have incapacitated and ill-equipped society to manage the co-occurring disasters at our doorstep. As Baldwin aptly exclaimed, "I imagine one of the reasons people cling to their hates so stubbornly is because they sense, once hate is gone, they will be forced to deal with pain." Reconciling past ills and trauma is not an easy task, and yet it is necessary for society to progress beyond the past. Dealing with intergenerational and historical trauma tends to be especially difficult for Americans, as the fear of being labeled racist outweighs the desire to abolish racist systems. But like the inter- connectedness embodied in Afrofuturism, our past colors our present, and if we let it, will dictate our future.

Research on the ecology of segregation appropriately highlights the profound impacts that past racially driven residential segregation can have on the ecology of urban ecosystems. Government-sponsored pro- grams like redlining—an exclusionary and discriminatory housing policy established by the Home Owners' Loan Corporation (HOLC)— categorized neighborhoods using color-coded maps to denote where American residents should live based on their race and wealth. Redlining was instituted from 1933 to 1968 for more than 230 US cities, and dic- tated patterns of investment, quality of municipal services, and greens- pace access. With the advent of the Fair Housing Act in 1968, which coincidentally was advanced by the actions of John Lewis and others, the system of redlining was abolished to make housing discrimination illegal. However, the vestiges of redlining are still felt on the ecological and health outcomes of our urban landscapes. Areas that were formerly redlined have greater air pollution, more intensified urban heat islands, reduced vegetation cover, and more frequent rates of ER visits and cancer. Though redlining policy was abolished in 1968—more than fifty years ago—cities are still experiencing the pernicious environmental disamenities that are linked to residential segregation from the 1930s to 1960s. Recent studies are also beginning to suggest that genetic diversity and species richness—two measures critical for measuring and sustain- ing biodiversity—are also being impacted by past racial inequities. The disparities even extend to how we perform our science. For instance, repeated evidence suggests community science platforms like iNatural- ist and eBird have tremendous gaps in species sightings in redlined neighborhoods, highlighting how urban residents are less likely to make natural observations in neighborhoods perceived as "dangerous" or "threatening." Coincidentally, societal biases are preventing us from doing our best science and making the necessary observations that help reconnect us to nature.

Emerging evidence on the legacy effects (i.e., the idea that past events and phenomena have long-lasting impacts on present social and ecological processes) driven by redlining shines a spotlight on the importance of interrogating past inequities as a driver of biological systems. Simultaneously, this and other social-ecological examples emphasize that the machinations of society impact biological systems. In other words, our actions toward each other—whether in the past or present—have pervasive consequences for all other organisms on this planet. The amplification of social and racial inequities can compromise ecological and climate integrity, making it an urgent priority to treat the root causes of societal ills as a requirement for building resilience to emerging climatic disasters. Yet again, undertones of Afrofuturism are applicable here, as the solution to addressing both ecological and societal ills requires time travel: the need to understand past events, their impact on the present, and potential projections into the future. This is not unlike the emerging field of climate futures, which provides a framework in which to envision the plausible structure of social and natural landscapes in response to several possible scenarios. Concurrently, our advancement as a species is principally hindered by our reticence to deal with the influence of past transgressions on our current and future ecological realities, which sabotages our ability to plan for and imagine any possible futures. If then, Afrofuturism and hope for a better future are the antidote to building resilient strategies that rescue the natural world, active ignorance of historical ills is the virus killing our link to the natural world.

Healing our connection to nature thus inherently means healing ourselves. To restore our connection to nature, it is a requirement to acknowledge that all human beings are part of nature. Doing so then advances our understanding of how we are connected to each other, as well as the other living and nonliving components of the planet. Perhaps as a sign of hope, discussions centered on the interconnectedness of people, the environment, and other living organisms has emerged repeatedly in the last few decades. For instance, the environmental justice (EJ) movement and culminating delegates to the First National People of Color Environmental Leadership Summit in October 1991 established the seventeen guiding principles that are critical to EJ praxis. The very first principle underscores the ecological unity with and interdependence of human civilization on nature. Moreover, the emergence of the One Health framework has gained considerable attention in organizing discourse on the interconnectedness of society, animals, and their shared environments. Ironically, the COVID-19 pandemic, which is a zoonotic pathogen, underscores how dependent our health, medical systems, and governance are on environmental health. In both activist (i.e., Environmental Justice) and academic (i.e., One Health) developments, undertones of Afrofuturism—hope for a more just future, community empowerment, and social-ecological unity—are pervasive throughout the DNA of each.

Still, considerable barriers exist for us implementing an Afrofuturist blueprint. Are we willing to be uncomfortable in the process, and divorce ourselves from the fear of pain to actually gain back our love of nature? Are we as a species willing to fight for all peoples' right to the natural world, even if we personally are not privy to the end result? An Afrofuturist philosophy may argue that the fight for a truly liberated future is worth the struggle, regardless of whether it comes into existence in our lifetime. If we see the multipronged threats to the natural world as opportunities to fight back our own fears and connect as a species, then perhaps our connection to nature is simply obscured, rather than lost. Trauma is exceptionally adept at obscuring our vision, and thus, envisioning an equitable and interconnected future is the act of healing.

This is all to say that humanity's collective journey back toward our love of nature is both exceptionally close and extraordinarily distant. Arguably, though a global majority of peoples acknowledge the significance of nature to humanity's well-being—especially teachings emanating from communities of color, Indigenous peoples, and the Global South—the status quo sustained by global capitalist ideals is antithetical to substantive progress in conserving nature. Specifically, racial capitalism (i.e., the concept that race is central in forming social and labor hierarchies in capitalist systems) treats nature, the environment, and peoples of color as expendable commodities to be owned and controlled, rather than being a liberated force deserving of life free from privatization. Simply put, racial capitalism and nature cannot coexist. The destruction of nature for profit and the widening of wealth inequality is not sustainable, and reforms or modifications to this system will not yield different results. If we are to make direct recommendations for how we regain our connection to nature, the first step will necessarily need to be the dismantling of a capitalist system built on the tenets of resource extraction and human exploitation. Creating such a paradigm shift will require supplanting an insidious and resilient system: white supremacy.

Afrofuturist narratives provide a compelling argument that suggests technological advancements through cultural ingenuity rather than maximizing economic profits is a possible future scenario, one of ecological coexistence rather than conquest. Creating space for various communities at local, regional, and global scales to weave in their traditional and experiential ways of knowing with conservation or climate solutions will pay untold dividends in shifting our viewpoint back toward nature. To give just one example: traditional Ghanaian structures use adobe mud blocks as the foundational building materials for homes, and the adobe serves as a brilliant architectural cooling agent in a tropical climate. Despite this ingenuity, British colonizers insisted on the use of cement blocks over the native adobe designs because they considered them more "sophisticated," despite cement being vastly inefficient at providing cooler microclimates. Increased construction of cement homes meant that more resources needed to be imported into Ghana and more residents would start using air conditioners to cool their homes, two factors that would contribute to global climate change. Ghanaian designers and architects like Joe Osae-Addo are now using more inno-native designs that reintroduce Indigenous textiles, materials, and engineering that are locally adapted, creating living spaces that rest in equilibrium with, rather than in opposition to, the environment. An Afrofuturist lens would elevate these types of stories to illustrate that a future more connected to nature means giving reverence to the wisdom of our ancestors' past innovations.

An Afrofuturist lens also provides us the space to think creatively about the future we want to see. Despite tremendous scientific advances in climate and environmental modeling projections, the future remains shrouded in uncertainty. It is hard to envision what future generations' experiences with the natural world will be, and this is especially the case when those future communities will face markedly different environmental conditions. However, this does not mean that we do not currently have the ability to plan for that future. Whether through science or science fiction, Afrofuturism unchains our consciousness to dream of what could be, and thus think about the steps necessary to make those dreams a reality. This is not unlike creating a meticulous plan that partitions what feels like an insurmountable goal into achievable, bite-sized deliverables that are less intimidating. Doing so allows us to lift ourselves out of climate "doomism" and creatively think about what is next, without knowing how that plan will ultimately turn out. This provides us a sense of agency in tackling a problem that can only be solved through collective planning and community engagement.

Finally, an Afrofuturist lens emphasizes that transformative societal and environmental change is simultaneously within our power and highly interconnected. The conservation and regeneration of the natural world is not simply maintaining genetic diversity, population abundance, and species interactions. These ecological metrics are similarly influenced by socioeconomic disparities, housing insecurity, aging infrastructure, and sociopolitical upheaval. Take, for instance, the recent cycles of droughts and flooding in the Bay Area, California. The summer's end in 2022 witnessed a massive die-off of aquatic organisms that shocked the entire Bay Area community. Investigations concluded that discharged, treated wastewater was contributing substantial amounts of nutrients into waterways that led to the explosion of algal blooms. These blooms blocked out sunlight and consumed inordinate amounts of oxygen, suffocating marine life. A few months later, in January 2023, several consecutive atmospheric rivers pummeled California, leading to flooding and landslides that incapacitated residents. Aging stormwater infrastructure contributed to additional runoff of nutrient-rich water into those same waterways that were just hammered by the summer die-off. Unfortunately, residents living in lower-income neighborhoods also experienced the brunt of these environmental burdens, with the greatest flooding occurring in the flats of the East Bay and the Bayview neighborhood in San Francisco.

This climate-driven one-two punch is but a single example amid a bevy of worldwide calamities, but these are not hopeless catastrophes. That's the thing: we already have the research and solutions to mitigate or prevent future environmental collapses like those mentioned above. Increasing social equity, centering environmental and climate justice initiatives, embedding empathy into collective governance, and greatly improving our educational systems are just a few of the verified solutions that will bolster our resilience to impending environmental changes. Operationalizing an Afrofuturist lens is not always in service of creating new recommendations or restating validated recommendations that have repeatedly been called out by Black, Indigenous, and other community practitioners. Afrofuturism is also the hope that we *can* change—that we can authentically implement proposed solutions once we become fully liberated, emancipated, and decolonized. We can build resilience into our cities and towns, suburbs, and rural neighborhoods with the important start of understanding that we *must* change our relationship with nature by seeing how we are connected to nature through each other. Luckily the ancestors and elders from the diaspora and beyond have etched that path of compassion and empathy in our collective memory. We simply need to regain our power to hear those past lessons for a better future.

In 2018's *Black Panther*, the late Chadwick Boseman's character T'Challa states that "Wakanda will no longer watch from the shadows. We must find a way to look after one another, as if we were one single tribe." The weight of that line underscores the need to unify all the world's peoples, find commonality in our perspectives, and collectively work toward building resilience. Chadwick, too, fought for a considerable number of just causes before his passing, while his memory lives forever on screen as the emblem of Black pride and empowerment. In 2022, when the sequel to the critically acclaimed movie was released, *Black Panther: Wakanda Forever* beautifully captured how tremendous loss, grief, and trauma can obscure our ability to see our shared struggle in others. Shuri, played by Letitia Wright, exclaims at the movie's conclusion that "vengeance has consumed us; we cannot let it consume our people." In this line, Shuri and Namor hold the generational trauma of their peoples and find commonality in their shared experiences with structural violence and brutalism. Both these lines—from *Black Panther* and *Wakanda Forever*—are exemplars of the beautiful struggle: the constant fight to live in a world antithetical to your survival, yet the willingness to devote that life to the success and service of others. May we learn to disconnect from an ideology of individualism to think more communally, and wholly integrate Black perspectives and Black people into our future. If the lessons from our ancestors have taught us nothing else, the time to act is always now, even if we never see the fruits of our labor.

15

Listening to Learn

NATURE'S HOT AND COLD EXTREMES

Joel Berger

THIRTY YEARS ago Archie Gawuseb killed a black rhino. He was one of numerous poachers operating in the Namib Desert. Shortly thereafter, I arrived to examine whether living rhinos that had been dehorned by the government could survive hornless. The radical tactic was designed to remove the incentives to poach.

I asked Archie the critical question—why, why he poached. "Life is unfair" was the answer.

———

The world has no shortage of injustice—pestilence and war, malnourishment and poverty, social dissonance. How is it possible to be an optimist when people kill for greed, or when the challenges of climate alteration appear to overwhelm; indeed as we witness even more assaults on biodiversity? Regardless of whether you are black or white in skin, privileged or disenfranchised, the planet is a mess. Beyond humans and our primate kin, rhinos are certainly not the sole taxa in trouble. Nor is a dazzling array of resplendent species, as we today pay homage to once-cherished landscapes.

Nature is the essence of life. For humans, the future for wild nature is not going to be pretty—despite a half billion Americans who hike, observe, recreate, and seek to just "be" in US protected areas annually, as similarly do some eight billion annual eco-tourists around the world. If we recoup curiosity, if we can inspire, and if we have a modicum of success in making clear why our current caustic practices will destroy us, the future can still be attractive. Progress will stem from our ability to listen, to learn.

Herein, I develop two narratives from the planet's raw edges. I showcase how two people lacking in both privilege and equity maintain hope in spite of all the disharmony. "Fairness" would never be the first word that comes to my mind. Yet, each individual loves wild nature, uses nature's bounties, and has had a strong hand in passing on generational knowledge. As is clear from other writers in this book, the assortment of examples and characters portrayed typify the requisite different approaches necessary to better conserve what we have left and restore what has been lost. I begin with Mr. Gawuseb in an African desert. I conclude at the opposite end of the world where a former reindeer herder named Freddy Goodhope Jr. helped me absorb his Arctic.

From Hope to Optimism in Namibia

Archie, a Damara whose native language involves a series of clicks and who has as a birth name Huie', was forced to have an appellation more befitting of the oppressive apartheid regime under which he was raised. He explained his decision to poach was not entirely based on social injustice. Nor was it based on greed. Poverty and an uncertain future were the central ingredients.

Archie's story is not unique. Though changing, many rural Africans are born in homes without running water and electricity and have little access to health care nor to formal education. Peril assumes many forms—food insecurity, disease-borne insects, venomous snakes. In Archie's case, there's also the occasional destruction by wild elephants to his family's most treasured bounty—pumpkins and corn.

Archie's chronicle, however, serves as an exemplar. No rags to riches here, but commitment, learning (sometimes by him, many times by me), and from hope to optimism. His narrative is a mix of causation blended with serendipity to conservation consequence. It's also a reflection of love for nature and the wild, a driving concern of this book.

For a young Archie, the nearest village, one with access to rudimentary amenities, was more than 270 kilometers away. Paved roads did not exist and autos unavailable; only multiday, torturous burro rides were available to reach the town of Khorixas. His father needed a permit to enter the white-ruled village, and he had to be gone by sunset.

Until independence from South Africa in 1989, Namibians were governed under the former apartheid regime whereby Archie witnessed the brutal twenty-six-year war. The backing here was global, covert and overt: advisers, operatives, or military, which emanated from Angola, Cuba, Russia, Zimbabwe, Great Britain, South Africa, Democratic Republic of the Congo (formerly Zaire), and the USA. Although Namibia's northern tier was a flashpoint, it was in the remote deserts where Archie grew up, protecting goat herds, and where wildlife flourished. The arid outposts average less than ten centimeters of annual precipitation, the people are few—ethnic Himba, Herero, and Damara—and eke out a meager living, mostly through agriculture and livestock.

The desert today still sustains remarkable diversity, much due to the rich fossil (subsurface) river systems. There are two-thousand-year-old *Welwitschia* plants. There are giraffes, elephants, rhinos, ostrich, oryx, and two species of zebras. There are lions, cheetahs, leopards, three species of hyenids, even the occasional wild dogs. Wildlife remains not because the environment is so hospitable—which it is not—nor because of low human density, which it is. It persists because the local people demanded it, take pride in it, and have created structures to maintain it. Archie's role, while modest, has been influential.

When I met Archie in 1991, he spoke five languages. He knew the landscape and its animals. As a tracker, he distinguished the hoof-prints of domestic goat from klipspringer, burro from zebra, and lion from leopard. He followed rhinos across pure rock and could tell the footprints of males from females based on patterns in sand. He'd walk thirty kilometers a day in 40°C heat. I was hopeful that if I observed and listened I could develop some of his skills. In addition to being the finest tracker I have ever known and among the keenest of natural historians, Archie is a convicted felon. In an unusual chain of judicial events, his court sentence was to benefit rhinos rather than to undergo incarceration. Because gang members had threatened Archie's life due to his court testimony, his assignment was to assist "the Americans"; oddly, it may have been for safekeeping because he'd be in remote field camps sequestered away from gang members who lacked the transport to find him.

My immediate goal was not focused on Archie's safety, but instead to discover whether the dehorning of rhinos had inimical effects on animal survivorship. The government-sanctioned dehorning program was enacted as an emergency measure to thwart incentives to kill rhinos for their highly valued horns. So, while I pursued the question "What is a rhino without a horn?" through Archie's tutelage I learned about wild nature.

Archie's story continues to unfold. In the intervening years since we've worked together, Archie developed "Elephant Song," which is his own tourist camp and which has attracted Israelis, Brits, Germans, the French, Brazilians, and Americans. He's received US government support, visited North America's Greater Yellowstone Ecosystem, and learned aerial census techniques and radio telemetry by on-the-ground and in-the-air training. He's helped international biologists study desert bats, giraffes, and rhinos, of course.

In 2021 Archie competed for the chairmanship of the Sesfontein Conservancy—a not-insignificant achievement as Namibia now has more than seventy conservancies that guide wildlife conservation and land-use decisions (fig. 20). The conservancy in Sesfontein covers about 2,500 km², which is more than twice the size of Grand Teton National Park in the US. Archie's son protects desert lions as a law enforcement ranger, and his daughter is a teacher in the village of his birth, Sesfontein. While in 1991 Archie judged his life to be unfair, suffice it to say, his unusual punishment coupled with opportunity encouraged powerful reform. In turn, he, I, and so many others maintain optimism, the result

FIGURE 20. Archie receiving recognition award from Sesfontein Conservancy.
Photo by the author.

of actionable conservation programs across scale and decades to appreciate wild animals and their arid lands under locally based directives.

A Reindeer Herder at the Other End of Earth

When one stares at a NASA image of Earth and looks north from the
Gobi Desert, or those far to the south such as the Namib or Chile's
Atacama, broad white bands stretch in all directions from the North
Pole. They used to be more extensive. But, it is these northern icy zones
that harbor the Arctic's largest land mammal, muskoxen. Here, precipitation in polar deserts mirrors that of the blistering Namib, with as little
as ten centimeters annually.

These distant and remote biomes were famously compared by
Charles Darwin in 1859: "Not until we reach the extreme confines of life,

in the Arctic regions or on the borders of an utter desert, will competition cease." While Darwin referred to population regulation, he implied that it is weather and neither competition between species nor predation that inevitably governed population sizes. His insights derive from an era when less than 1.5 billion people inhabited Earth. Industrial footprints were embryonic and much of nature played out without much, if any, human interference—at least at the planetary edges.

Today, we've exceeded eight billion. Every landscape and seascape have been affected by our unmitigated thirst for resources. Take muskoxen (fig. 21). Once circumpolar, most of their contemporary change occurred due to us humans. In the 1840–1860s, it was whalers of the northern Bering and Chukchi Seas who introduced guns to Alaska's native coastal people. By 1890, muskoxen were extinct in Alaska, and in serious declines in Canada. Conservation measures enacted in the 1930s returned muskoxen to Alaska. By the 1970s and 1980s, notably through aggressive state management, they occupied most suitable Arctic habitats. Decisionmakers in Juneau, and later, Washington, DC, were initially responsible for the financial backing to rewild a part of the landscape—hence, an unwitting victory for an unheralded species of little notoriety to most Americans, which included, of course, Native Alaskans.

Muskoxen once roamed a grassy tundra with woolly mammoths. They outlived the elephantine goliaths by some ten thousand years in Alaska, although mammoths managed to persist on a distant Siberian Island (Wrangel) as the first Egyptian pyramids were being built. Although climate is clearly implicated in the demise of mammoths, the impacts of changing weather on muskoxen has not been especially obvious.

My work on the shaggy behemoths began in 2007. I aimed to understand how the loss of ice and other aspects of rapid climate change might conspire to challenge muskoxen. Without such knowledge, it was obvious that long- or short-term conservation could not be formulated. This meant knowing about species interactions. Perhaps parasites or diseases are being propelled northward with warming temperatures or changes in host species distribution. Also, with grizzly bears progressively moving to deglaciated areas and with more polar bears on land, it's also possible that predator-prey relationships are changing.

FIGURE 21. Muskoxen group in flight (from left to right are adult male, adult female, yearling, and juvenile male). Photo by the author.

I felt my efforts would be enlightened by learning about the species through the eyes of Alaskan natives—people who go by the name of Inupiat or Eskimo—as their ancestors colonized Beringia and harvested, in addition to muskoxen, horses, mammoths, and giant bison. But, as I had discovered in Namibia, studies of biology and ecology are not sanitized nor can they be disentangled without an indebtedness to the social and cultural milieu of local people. In other words, understanding lessons of history and justice from the people with deep tendrils to the land or coast serve centrally in conservation effectiveness. I wasn't prepared for what would follow on my first field day, one in which I scouted habitat and animals from the air.

"Why do you want to fuck around with them muskox?" was the first phrase I heard as I stepped down off the bush plane in Kotzebue, on the shores of the Chukchi Sea. Willie Goodwin, the ex-mayor and respected Inupiat elder of this Native village, offered the remarkable greeting. The predominant assumption among local hunters was that muskoxen compete with caribou, the latter being far more numerous, less dangerous,

FIGURE 22. Fred Goodhope Jr. Photo by the author.

and valued as the winter mainstay for residents of Arctic Alaska. Willie elaborated on his despair about Alaska's former conservation decision. "No one ever bothered to ask us, not when muskoxen were brought back to Alaska, and not now, whether the people had ever wanted them. It was our land. No one asked."

Although Willie had been a subsistence hunter and later a valued friend, I sought a local native guide with, perhaps, deeper knowledge about muskoxen and savvy about Arctic survival. That man was Freddy Goodhope Jr., previously a reindeer herder, a pilot, and a dynamite naturalist and mechanic (fig. 22).

Just as Archie Gawuseb knew his desert environs like the back of his hand, Freddy could find his cabin ninety kilometers up the tortuous coast in an iced fog or total whiteout without GPS. He knew safety, having reluctantly killed both polar and grizzly bears. He knew kindness, and shared his meager wares with me, as I did equally with him. For nine years we did winter work together.

Like kin, Freddy and I knew without words when there was danger, where to cross a frozen river, or how to navigate soft snow in a deep canyon. We read muskoxen behavior, knowing when to approach or to watch from afar. We mused about our cultural differences. Freddy did not like to see a wolf alive. I did. He liked to shoot Arctic hares. I wouldn't. I did not grow up nor winter in the Arctic but enjoyed the friendlier beaches of Los Angeles. Grizzly bear and seal skins hung from drying racks in his home village, Shishmaref. We agreed to disagree on some practices. Yet, the respect was mutual. Through Freddy, I began to understand Arctic species, challenges, and rapidly changing sea- and landscapes.

Freddy is an admitted throwback to an earlier era when ingenuity, subsistence, and, hence, survival were dictated by the landscape and not by access to a grocery store. As a pilot, he once crashed a plane just offshore into the Chukchi Sea when counting his reindeer herd. He's used his snow machine to rescue people lost in the Arctic. One winter Freddy brought in thirty-five caribou to help feed the people of Shishmaref who were unable to hunt.

Subsistence is still critical in remote pockets where 190 Alaskan villages have no road access, and can be reached only by plane, boat, dogsled, or snowmobile (fig. 23). While many in these outposts understand seasonal cycles of nature and use the resources, others remain less interested and are drawn to the larger population centers in places like Nome, Fairbanks, or Anchorage. When I ask Freddy's children if they are attentive to changes in land, ice, and wildlife, they say yes. But they happily live with more convenient amenities than in remote locales such as Shishmaref. Consequently, digital screens are the more active mode by which to enjoy nature.

FIGURE 23. Freddy, snowmachines, and sleds in a stiff wind—Arctic Beringia. Sleds carry ~ two hundred liters of fuel and camping supplies. Photo by the author.

Freddy still resides in Shishmaref. He shares his knowledge of nature's ways when people wish to listen. Reindeer herding in Alaska has not been profitable for decades. Remnants of Freddy's dilapidated reindeer fence remain visible ninety kilometers up the Chukchi coastline, where occasional wild caribou pass by. Freddy's self-built home in "Shish" burned to the ground in winter 2021. All belongings were lost. Still, Freddy, the elder, does limited guiding.

Lessons Learned by an Old White Guy

It's now thirty years since my first visit to Sesfontein. As a seasoned and timeworn white guy, it would be easy to sink into deep cynicism, given our world's extraordinary challenges. The COVID-19 maelstrom and its associated mutations have excised untold suffering, as will future pandemics. Greed and short-term blindness foment climate cataclysm, and, sadly, many elected decision-makers choose inaction over attempting to thwart or reverse the ravaging of global landscapes, the spread of micro-plastics, and the loss of clean air and water.

Looking forward, a path should not be one of utter despair. It should shine with optimism. I believe this emphatically—such a perspective is anything but Pollyannaish. Many of our inspirations stem from sparks of successful conservation at home, in towns, cities, and certainly at some broad scales in the less trammeled natural world. While it's fashionable today to classify early promulgaters of conservation and people of my ilk as old white guys, curiosity and commitment across lifetimes led to today's victories—notables like John Muir, Theodore Roosevelt, William Hornaday, Aldo Leopold, George Schaller, and Michael Soulé. Similarly progressive, yet facing the more onerous road of being a female, are Rachel Carson, Jane Goodall, Wangari Maathai, Ruth Buendía, and Sylvia Earle. Inspirational writers of the ilk of Terry Tempest Williams, Elizabeth Kolbert, and certainly the curmudgeonly desert rat, Ed Abbey, have catapulted earth's living support systems and nature to a larger public eye. David Quammen has fused science with narrative in his brilliant reads while David Attenborough reaches more through film. Canadian-born musicians Joni Mitchell and Neil Young have reached millions, as well, through passionate lyrics about what we lose as Earth's remarkably diverse life-forms slide into obscurity.

But it has been the nameless advocates who continue to stimulate curiosity and instill future values. Beyond the visualizations of artists, cinematographers, and photographers are journalists, educators and students, naturalists and guides. Birders, hunters and fishers, and recreationists refuse to relax their guard. Indeed, accumulated knowledge of changes in biosphere have bolstered movements to mitigate climate degradation, including ways to reduce losses of global ice and additional protections for cold-adapted species. While white guys—some old, some not—initially played strong roles in raising the conservation bar and where such privilege offered a modicum of entitlement, inclusiveness of all people is finally emboldening the conservation pursuit.

My narrative of two indigenous figures depicts poignant, though very different, relationships with the land they grew up on. Notably, Archie and

Freddy's profound understanding of the natural world emanate from places that still today are of low human density and where wildlife extinctions are lacking. These are not the conditions of most of the world's nearly eight billion people, nor will most ever feel or know the extreme worlds of Archie and Freddy.

Today's onslaught on nature occurs for countless reasons but not least among them is a lack of exposure to nature, ignorance, or the belief that we know all and can circumvent the services nature provides. For many people it will be a deeply myopic concentration on the immediate world without serious reflection on the consequences of today's actions. Harry Greene in his brilliant book, *Tracks and Shadows*, recounts how it's never too late to connect to nature at an individual level. At times, all that might be needed for that deep breath is a walk in a desert or a forest, a boat ride across a lake, a visit to a zoo, or touching a dinosaur bone in a museum. Yet, many never even have such opportunities. More than half the disenfranchised youth of Los Angeles or in Denver have never been to the ocean or the mountains respectively.

In bizarre ways, the digital world can help connect with nature (as Susan Clayton and Bill Adams consider in their contributions to this book). Perhaps Fat Bear Week—about bears gorging themselves to fatten in preparation for winter hibernation—is all it will take to motivate a viewer to visit Alaska, to watch bears there, or to catch their own salmon much as bears there do. Perhaps it's that visit to a national park, or an active classroom experience of learning how to track wild animals in snow, or one on hunting or identifying birds, or watching a bee finding nectar in one of your garden's flowers.

I began this essay with flashbacks to reflect future hope. It's not because of Freddy and Archie per se but because ubiquitous are the people of character—kind, generous, and able to inspire.

Possibly this is nowhere clearer than the remarkable path taken by a boy who grew up in a farming family in Senegal. His name is Baba. His father worked rurally, raising cattle and selling peanuts. Baba helped. Eventually, he attended college and became a forest engineer. In 1967 he became the Senegalese director general of Water and Forestry. Baba

Dioum is best known for a quote in his 1968 speech to the IUCN (International Union for Conservation of Nature) delivered in India: "In the end we will conserve only what we love; we will love only what we understand; and we will understand only what we are taught." The challenge remains, of course—exposure, listening, and learning.

Steady progress to restore wild nature and its species continues nonetheless, covertly, overtly, and through happenstance. A case in point for the latter is the reclamation of land. Parts of the arid central prairies in the US are depopulating as rural Oklahomans, Kansans, and Nebraskans emigrate from their no longer productive farmland. More striking is the continental scale and makeover for South America. By 2050, 85 to 90 percent of the populace will be urban. That means in just two and a half decades much of the countryside will be of low human density. Already parts of Patagonia are being reclaimed by guanacos as sheep ranching becomes less profitable and the land loses its domestic stock. Pumas, in turn, and without past levels of persecution, are recolonizing these regions. Governments, nonprofits, and a new wave of educators invest financially to seek brighter economic futures. Urban and rural dwellers alike—with conservation sympathies—desire more protection of additional focal zones where nature is less impaired by industrial complexes.

Societal attitudes toward nature do change, and species protection and expansions occur as a consequence. There are many reasons that hope can turn to optimism. Consider Rwanda, the second-most densely populated country in Africa. Mountain gorillas continue to increase in population size. Alternatively, other species are extirpated in the wild but do rebound.

Growing up in California, I saw my first and only wild condor in the late 1970s some forty miles north of Los Angeles. Within a decade they were gone. Today, however, because of captive propagation and release back into the wild, there are about 270 distributed among Mexico and the US, including the states of California, Arizona, and Utah. For grizzly bears the story is similar, though details differ. In the 1970s, the last population in the lower US numbered maybe three hundred, almost all in Yellowstone National Park. Today, due to protections beyond the park that the public supported, there are seven hundred to eight hundred there and another one thousand across western Montana into Idaho. Same story for wolves. While each of these three species and many more face enormous challenges due to climate, more people on their landscapes, and frenzied intolerance, we are learning to live with these species, and new generations continue to seek ways to be kinder to nature.

We don't know for how long. If we do not appreciate nature, who will be the caretakers? Who will speak up? Without curiosity, minds grow dim. Demographically, too, the challenge remains steep. Momentum must come not only from rural sectors where families persist with more nature, but increasingly from our urban landscapes where a majority of us reside and where a growing disconnect governs our relationship to nature.

Rekindling and maintaining a love of the wild on an increasingly anthropogenic planet requires new methods. In his book *Our Wild Calling* (2020), journalist Richard Louv argues for nature-based learning in which our K-12 students develop empathy and curiosity, and experience animals through introduction to them in the classroom or through interactions with domestics—a bit along the lines of a 4-H model. Empathy for our wild kin also develops through celebration of charismatic individual animals, such as the remarkable mountain lion P-22 who roamed the hills of Hollywood, grizzly bear 399 who delights Grand Teton National Park visitors, and Yellowstone's O-6, deemed the "Most Famous Wolf in the World." Time spent observing other species builds empathy, broader curiosity, and a place in the heart for all of life's connectedness.

A love for nature, just as it has in the past, now, and into the future, arises in many ways, from serendipity, through exposure, and some by planning. It really does matter if people continue to care and to act. The stories of Freddy and Archie reveal vitality, their reliance on the land, and an appreciation of the species and the associated wild. This needs to be us—our hearts, our minds, whether close at home or from afar. In the end, the question rephrased for Archie due to the poaching incident was "Is life fair?" No; of course equity is not spread evenly. But, if we fail to engage people from every spot on Earth to partake in protecting the environment and its consequent biodiversity, life as we know it will only grow sadder.

16

When Natural History Brings Us to Our Senses

Gary Paul Nabhan

Fieldwork has also been contemplative, inspiring me to pay attention and live more fully.

—HARRY GREENE, *TRACKS AND SHADOWS*

I WAS WADING waist-deep in the muddy, swirling waters of a flash flood as it slowed and settled into a *charco* stock tank, a kind of rainwater retention reservoir common throughout the desert borderlands, in this case one that was all of a sudden teeming with toads in what looked and smelled like the mythical primordial soup of (pro)creation.

They had all been hibernating, estivating, or contemplating deep in the sand until the first thunderstorms of the summer came and beckoned them to come up for some ozone-charged air and a probiotic drink.

I was groping my way through the dark, guided only by the light of my own headlamp, for my five friends had scattered in the pursuit of other amphibians that had dug themselves out of the desert ground. Those amphibians were moved to emerge by lightning or by thunder; by thunder or by moisture; by moisture or by the pheromonal fragrance of tens of thousands of the potential mates. Those mates in waiting had rafted the waves of the surging flash flood, until they had all converged back at their birthing grounds.

They were seizing the moment.

I had arrived just an hour or so before, after a number of us received an alert to their whereabouts from a herpetologist who had embraced the notion that for critters adapted to the desert's uncertainties, being on time for an orgy was everything.

They had no time to lose, for they come together only once in a blue moon.

I had crawled out of my bed still half-asleep and barely half-dressed, jumped into my beat-up pickup truck called Old Paint and converged with my friends where the pavement ends, so that we could follow one another's tracks as we fishtailed down mud-rutted roads to descend upon this frenzy of amplexus and fertilization.

They paid me no mind as I struggled to sort out the toads into their respective species. *Bufo alvarius, Scaphiopus couchii,* and *Spea hammondii* had all floated down the same flood-filled arroyos to attend this midsummer night's dream.

As a botanist on my maiden voyage of sailing into a sea of toads, I was rather inept at this sorting, compared to my five more herpetologically seasoned friends, who seemed to be able to distinguish the species from one another in less time than it would take me to sing a bar of "Froggie Went a-Courtin'."

These toads were drenched in mud, so that the diagnostic traits used to distinguish one taxon from the next in illustrious field guides were obscured, such that much of the prepping I had undertaken prior to arriving at the pond went down the proverbial drain.

I could not tell which of spadefoot toads that I grabbed and pulled up into the beam of the headlamp had the sickle-shaped spade on its hind foot nuptial pad that was characteristic of *Scaphiopus*, and which had the less angular and roundish spade diagnostic of *Spea*. As a sleepy-headed novice, I was also having a hard time seeing whether the toad in my hand had eyes with the vertical pupils of spadefoots, or eyes with the horizontal pupils of the puffier Sonoran Desert toads of the genus *Bufo*.

To make matters worse, all the toads were so hyped up and horny that they were no longer mating in twos, but were humping each other in long chains—hook-ups of multiple species—that stretched out like an accordion when I tried to lift them up out of the waters.

How could I discern the relative abundance of the three different toads when the males showed no concern for whether their nuptial pads were grasping females of their own species, or another?

And yet, my colleagues seemed to be able to distinguish one species from another, even though I was failing miserably at that same task . . .

Then, rather suddenly, I came to my senses. As any proficient naturalist learns to do, we *all* must come to our senses.

The visual diagnostics in the field guide were simply not the cues to use in such a setting. In order to solve this riddle of identity and to record the proper ratio of the three kinds of toads that had come to celebrate together that night, I needed to shift my synesthetic balance.

I had to *listen*. And when I did, what I heard was that some of the toads were letting loose with cries that reminded me of the bleating of Churro sheep, while the Hammond's spadefoots nearby sounded out a short metallic trill that was like the scraping of a ratchet. In contrast, the Sonoran Desert toads burst into song with a low-pitched screech a bit like a ferry boat whistle.

"Hey *compadre*!" I yelled out to my closest friend, "Which of the three toad species sings '*Whaah! Whaah! Whaaaaah!*' like a lamb looking for its mother?"

"That's the Couch's spadefoot: '*Whaah! Whaah! Whaaaaah!*'"

I realized that was in fact the sound being made by the very toad that was clinging to my gloved hand as if he was ready to mate with it.

I had to *sniff* out the peculiar fragrances before me. And when I did, I was able to ask my friends a question that I could not have articulated before that wondrous night:

"And which one has that thick, rutty fragrance reminiscent of roasted peanuts?"

"Some say both spadefoots do," one friend replied.

Another offered, "I pick up that nutty scent more from Couch's spadefoot, the midsized one. For Hammond's spadefoot, I perceive a hint of garlic, but not everyone agrees with me. It's like the dilemma with those who think ponderosa pines all smell like butterscotch, versus those who pick up notes of vanilla in some ponderosas more than others . . ."

"And what about the big dude, Sapo Grande?" I asked. "You know, he looks all shiny and sweaty like those shellacked toads dressed up in Mariachi regalia playing trumpets, violins, and stand-up basses . . . like they sell in curios stores in border towns . . . if you sniff him up close, does he have a different fragrance?"

One of my friends wagged his finger at me.

"I don't recommend that you do *any* sniffing or licking of those big, fat Sonoran Desert toads. They secrete a milky ooze from the glands on their skin that is loaded with bufo-toxins that will either get you high or make you die."

It sounded as though my compadre had some personal skin in the game with regard to that binary set of options. Perhaps he was concerned for my welfare, but I suspect a baser motive: it seemed that he did not want me to get mixed up in toad-licking or toad-sniffing pursuits until after I helped him get the toad ratio tallies right that night.

That was all for the better, for it gave me enough time to contemplate how becoming a naturalist does not just happen through *eye-to-brain* cognitive processes, but comes instead through the exercise of all of our senses: *smell, texture, taste, sight,* and *sound.* And when all five senses play a role in our field practice, it seems that our *hearts* and *souls* get engaged as well.

And by becoming more attentive through all of our senses, we become more quintessentially human, for those are the very vehicles that made us more responsive to particular opportunities in our pluriverse that sent us on the evolutionary trajectory that we have taken up until this most recent stitch in time.

———

Ten years later, around 1990, flash floods from another summer thunderstorm appeared to be moving right toward me. I was living on the border at that time, in Organ Pipe Cactus National Monument.

One summer day, I was waiting at my home for a journalist from Dallas to appear to join me for a brief field trip to visit the largest ironwood tree just across the border, less than ten miles into Mexico. An hour after her projected arrival time, I had begun to worry whether she would beat the storm, or at least get there in time to get across all the soon-to-be flooding watercourses that would be blocking the road to the ancient tree.

The sky was purple, the wind had kicked up dust, and a sprinkle of light rain had suddenly begun to spray down on the thirsty desert vegetation.

A desperate knock at the door alerted me that the journalist had arrived.

"Can I come in?" she cried. "As soon as I parked out front, it began to rain cats and dogs. Wow, what is that awesome smell?"

The desert always smells like rain in moments such as this, I thought to myself. But what I said to her was not so poetic, since I had been contacted to be her science geek for the day.

"Well, the wet desert earth picks up the pleasant smell of rotting soil microbes, with a fragrance profile called *geosmin*. But the prevailing smell of the moment is that of the volatile oils released by the leaves of creosote bushes—a bunch of phenols and terpenes . . ."

I stopped talking for a moment to listen to thunder, and to the thunderous pounding of rain on our hot tin roof.

"Well, there . . . how interesting . . . But do you think we can get out to that gi-normous ironwood tree you told me about?"

"Was the wash running very hard when you crossed over the ford between here and the Visitor's Center?"

"It was a trickle then . . . but it wasn't raining then as hard as it is now . . ."

"Let me call over to the park's switchboard at the Visitor's Center."

I must have been frowning as I listened to the operator inform me that all roads out of the park—and over to the Visitor's Center—had been closed due to flash floods.

"Oh dear! What do we do now? I have to come out of here by the end of the day with a desert story for our magazine . . ."

I glanced over at this journalist in distress. What could I do to help her?

Suddenly I heard a familiar sound rising up in the desert: "*Whaah! Whaah! Whaaaaah!*"

"Could you excuse me for a moment? I need to check something out across the road, but I have an idea for what we can do. By the way, do you have any shorts or T-shirts you could change into?"

The rain was letting up, but the driveway and road in front of the house were still covered with a two-inch-thick sheen of sheet flow. I had on Tevas, so the water depth didn't much matter. As I walked across the road toward the wash, I could not only hear the surge of spadefoot voices; I could also catch whiffs of roasted peanuts wafting through the air.

When I reached the banks of the usually dry arroyos, I could see dozens of heads of spadefoots bobbing by, along with leaves, twigs, froth, dirty foam, flotsam, and jetsam. The water depth was at least three feet—maybe five at most—but not enough to drown any swimmers.

I ran back to the house, where my visitor had already changed into her running gear.

"What's up?" she wondered aloud.

"If you are up for it, we can tube down a small arroyo and follow the floods down to a borrow pit a mile away or so. It is not just filling up with floodwaters; it will have a congregation of desert toads that will provide great choir music for us for the next couple of hours. I'd suggest leaving your phone and camera behind for now . . . maybe we can go back for photos a bit later . . ."

"I'm up for it!"

I did not need to change clothes or shoes, so I ran to the garage to get two old inner tubes that were still filled with air. Within a couple minutes, we had plunged into the flooded arroyo with our tubes, and were moving along at about three miles an hour, using our feet to bounce off the banks, or boulders or tree trunks whenever needed. It was not hard to navigate past the hazards in most cases, so we could devote most of our attention to the toads bobbing along beside us.

Sometimes it takes a fluke encounter with another species—one where we meet it on its own terms, at its own speed—to remind us just how enchanted the world can be. We were doing basically what they were doing—sailing along on the frothy waves of floodwaters while storm clouds, lightning, and rainbows flashed above us in an equally turbulent sky.

The sound of rushing waters saturated our ears. The smell of roasted peanuts overwhelmed our noses—no hint of garlic was evident in this wash. We moved along slowly, peaceably, for there was no real race with the toads to get to the borrow pit before they did.

But once the floods fanned out in the shallow gravel quarry of the borrow pit, the soundscape shifted dramatically. It was as if we had front-row seats—tubes really—to the Vienna Boys Choir of Spade-foots. There were so many voices rising and resonating off the stony borrow-pit banks that it felt as though we had been invited to a private performance of the world's best throat singers.

And we had. We were in ecstasy. There is no other word to describe the joys of being among such mud-mottled singers, such dogged swim-mers, such hedonistic mates on their finest day of the year.

So much mud. So much music. So much unabashed mating.

———

Twenty years to the month after that moment in ecstasy, I found myself driving a van northward through Organ Pipe Cactus National Monu-ment after spending the Fourth of July weekend at a midsummer New Year's celebration of the Comcaac Indian community on the Sonoran coast. We had some delays at the border, and there was extra congestion due to all the vehicles present for the early phases of President Trump's mandated border wall construction.

Once we broke free of the traffic surrounding the Lukeville Port of Entry, I went into a reverie about my good luck in being present for the spadefoot orgy two decades before.

I scanned the desert landscape out past the roadside, trying to recognize where that flooded arroyo must have poured into the pond that filled the borrow pit that day . . .

Then suddenly, it dawned on me that something altogether unprecedented was happening to some fifty sizeable arroyos in the park that had historically carried spadefoots across the border to mate in Mexico, as the one with a terminus in the borrow pit had once done before that quarry had been bulldozed. Trump's wall would block—for good?—the flows of both floodwaters and spadefoot toads that had long traversed the international boundary line.

Just as I began to silently grieve over this newly recognized fact, I spotted something moving in the shadows of a tree on the other side of the paved road running north from Mexico.

As I passed the ironwood tree that shaded the roadside, I saw that it was a young, dark-haired woman and a small child with a baseball cap snuggled up against the trunk of the ironwood.

They were no doubt sitting in its shade to escape the glaring sun on that July day, as ground temperatures were already reaching upward of 120 Fahrenheit, although it was barely past 11:00 a.m.

"*Whoooa*! Laurie, did you see them?"

"See who?" she asked, looking up from her prayer book.

"A young mother and a child . . . maybe refugees crossing the desert . . . who else would be out in this heat, miles from any building?"

"Go back. Turn around as quickly as you can. *Go back.* They may be stranded and need medical help."

The professional nurse in her put down the prayer book and readied herself for action.

I circled back and parked on the roadside thirty feet south of where they sat, huddled together. Before I could pull out the key and take off my seat belt, Laurie was out of the van, kneeling in front of the mother and child, speaking to them in Spanish.

When she turned toward me to offer a synopsis of their conversation, Laurie was visibly distressed.

They were Central Americans escaping violence back home.

A *"coyote"* had dumped them out at an arroyo where they could easily crawl under the border fence somewhere east of here at dusk.

They had wandered for ten hours, lost in the desert, unable to see any lights to the north of them.

Their human trafficker had recommended that when they saw the lights of Metro Phoenix, they should just walk over in that direction.

"What the hell?" I blurted out. "What kind of evil soul would tell them that? Phoenix is more than a hundred miles away. Even the closest town north of here—why—is at least twenty-five miles away. There is no available surface water in between here and there. They would die if they had kept walking northward through the desert rather than coming over to the highway."

"Well, they are already noticeably dehydrated. No water left, no food. We need to get them some medical help. I can do more diagnostics, but if they are approaching heat exhaustion, they need to get to a clinic . . ."

I grimaced.

"Our only real option is to take them up to the Visitor's Center three miles or so north of here, and ask for help from the rangers. We can't drive them any further north than that . . . there's a Border Patrol checkpoint up another ten to fifteen miles, and we won't be allowed to have them in the van past that point . . ."

We helped the mother and child up into the van, gave them our canteens. Once I turned the air conditioner on for them, I began to drive slowly northward, while Laurie explained to them their rights and their legal options.

As I glanced across the road in the direction of where I had seen the spadefoot toads complete their safe passage into their mating pool, I realized that families of human refugees such as this one will not ever cross the border into this desert wilderness again, if the wall construction is ever completed. In fact, at least fifty watercourses significant to toad movements between Organ Pipe Cactus National Monument and Mexico would soon be blocked by the new wall.

The wall would impact far more than the movements of desert toads.

I had come to my senses with the help of some human friends and mating toads forty years before the day I met our country's "refugee problem" face-to-face, in the form of a young woman and her adolescent son.

I could smell their sweat.

I could feel the heat emanating off their bodies.

I could hear the heartbreak in her voice as the mother began to cry, realizing that they would likely be deported back into Mexico. She cried out that she could not bear to return to Mexico with her boy, because something terrible had already happened to her while they were stuck for months in Mexico City.

I could see the fear in her eyes, but her son did not yet understand that they had more trials to weather in the coming weeks. He was happily playing with a stuffed animal I had gifted him, a wide-eyed desert bighorn sheep.

"Be strong and wise like this creature," I told him just before we parted company. "He can teach you how to survive all the uncertainty out here in this desert."

Afterword

A PART OR APART: OUGHT NATURE LOVERS EVER WEAR FUR?

Harry W. Greene

I'll have no rebirth, but I will be in the bark of trees and a breath of air,
I'll not be in a church, but in the cells of the leaves and maybe I'll see
you there.

—RITA HOSKING, "RESURRECTION"

I AM HUMBLED to have inspired other naturalists, and for controversial
topics often prefer "show" over "tell." Accordingly, tasked with providing closure to *The Heart of the Wild*, I offer some personal journey to
complement the previous essays. Their authors include dear friends,
like coeditors Ben Minteer and Jonathan Losos, as well as scholars I
still hope to meet, like Eileen Crist and Chris Schell. Among them are
people with whom I usually agree, others not so much, and we've all
roamed locales that were wild by some standards. I've hefted my
backpack through Utah's Buckskin Gulch, among the world's longest
slot canyons, and trudged without guide or porter over Peru's thirteen-
thousand-foot-high Dead Woman's Pass on the trail to Machu Pichu.
I've hiked in rainforests with jaguars and bushmasters, waded in swamps

with caimans and anacondas. My snake-obsessed path often has paralleled Rick Shine's, yet fear of extreme cold, never mind the appeal of polar bears, means I'll not walk in fellow ethologist Joel Berger's footsteps.

And at every turn of life's pages, lacking the soul of a philosopher-poet, I cherish the words of Kathleen Dean Moore and others of her ilk for lighting my way.

The authors of this volume, despite varied perspectives, care about individual organisms, the fate of species, and wild places on a rapidly changing Earth. Maybe one day we'll see each other, as singer Rita Hosking imagines, in the bark of trees and the cells of leaves. Meanwhile, our essays won't provide the last words on wildness and wilderness, but I hope they encourage others to engage ever more deeply in this vital conversation. Nuance often lurks in emotionally charged controversies, and perhaps these reflections from an academic gone a bit feral can illuminate some of this one's nooks and crannies.

———

The violent deaths of two Texas carnivores, separated by more than half a century, are linked through what until recently felt like unrelated events. As a teenager looking for snakes on a friend's ranch, I'd cringed when his hired hand shot a coyote, one of its paws mangled in the rusty jaws of an old-fashioned jump trap. The men's tales of livestock loss smacked more of hatred than facts. Their victim's long snout, pointy ears, and bright eyes reminded me of Sunny, our family's first pooch. Worse yet, only years later, engaged in graduate studies of behavior, I realized the snarling animal's curled back lips signaled fear rather than aggression. And as a nature lover since childhood, watching that coyote perish, I couldn't have imagined that farther down life's road it'd be me aiming at another doomed predator. I couldn't have foreseen how an individual raccoon's demise might save other animals, or that snippets of verse would frame all of this as a broader dilemma: What does it mean to ethically, sustainably participate in nature?

After graduating from college and a stint as an army medic, I earned a PhD supervised by Gordon Burghardt at the University of Tennessee.

FIGURE 24. Sea otter using kelp blades as an anchor while eating crabs in Monterey Bay, California. Unprotected humans confined to these 55°F waters would die within hours. Image by Kathy West © 2018. Used with permission.

By the mid-1970s Gordon was already an expert on bear and reptile behavior, and fast becoming a major player in studies of nonhuman cognition and sentience. For me, thus mentored, there followed two decades as a California professor, during which a favorite diversion was observing sea otters in the surf off Point Lobos, grateful those largest, most charismatic of weasels were no longer harvested for their lush pelts (fig. 24). Hiking with my Berkeley students, I was privileged to search for salamanders among redwoods and snakes beneath Joshua trees, to pick through owl pellets in alpine meadows and bobcat scats on desert bajadas. And out on the West Coast, where John Muir founded the Sierra Club and Ansel Adams, subject of Ben's essay, helped bring that group into prominence, I often resonated with these lines from Robinson Jeffers's poem "The Answer":

Organic wholeness of life and things,
the divine beauty of the universe.
Love that, not man apart from that,
or else you will share man's pitiful confusions,
or drown in despair when his days darken.

Fast-forward thirty-odd years and an easy two-hour drive from the solitude of my Hill Country cabin. After a couple of more decades professoring in New York, I had moved to a ranch in Texas and was buying boots in an Austin outdoor co-op. What looked like fur caught my eye because, having veered late-career into anthropology, I'd just been reading about the history and merits of using animal skins for clothing. This parka's label falsely claimed its fake ruff is more effective than the real deal, adding, "We source high quality faux furs in a wide selection of designer colors and fashionable patterns that look as good as they feel." No organic options graced the chrome coat racks, no eye-catching signage credited ancient origins of winter apparel, yet dozens of quart water bottles by checkout proclaimed, "A part, not apart." Below each slogan a forest green coyote with red-dot eyes lifted its open mouth against a butterscotch moon, but the only audibles around this pack of plastic canteens came from a couple arguing about the price of a fancy kayak.

Seriously, designer colors and fashionable patterns? And what's with that fucking green coyote? Abruptly, I recalled that in the 1980s, San Francisco–based Friends of the Earth titled its newsletter "Not Man Apart."

Outside in the parking lot, exhaust fumes and traffic clatter, the ubiquitous chirping of urban grackles. I strolled out of the store with an old white guy's apologetic nod to the women excluded by Jeffers's words, then wondered what can "not apart" mean for anyone on today's greed-ravaged Earth? Ought shoppers have the choice to wear fur? Could Indigenous traditions of ecological kinship help rewild our lives and the planet? And recalling the editors' introduction, how can we keep a love of wild nature alive on a human-controlled Earth? These questions circle around a bothersome but often unstated truth, that other organisms aren't just hanging around out there admiring one another—or to co-opt another catchphrase popular among my fellow environmentalists, no other species has ever left only footprints and taken only photos.

———

Biologists unpack Jeffers's "organic wholeness" with specific terms. *Ecology* is the science of populations, communities, ecosystems, and landscapes. A *population* includes all the individuals of one species in a place, its numbers determined by births, deaths, and how many can exist. A *community* consists of the species in a habitat, say a meadow or pond, and is characterized by relationships like predation, competition, and mutualisms. An *ecosystem* includes communities, along with sunlight, water, and other abiotic ingredients with which they interact. Ecosystems are defined by relationships between organisms and nonliving nature, such as carbon cycling and energy flow via herbivory, carnivory, and scavenging. *Landscapes* include more than one habitat and are best visualized from the air. Landscape ecologists study habitat patches— shapes, sizes, connectedness—in relation to the populations, communities, and ecosystems within them.

Needless to say, a few dozen words of backstory cannot fully do justice to nature's richness, but they exemplify the multifaceted ways by which organisms connect to each other and their environments, including— inevitably, inexorably—us. If this still feels too abstract, pick a plant and an animal, say a claret cup cactus and a jaguar, then envision ways in which each might relate to other individuals of its same species, to other species in a community, to energy and nutrients in its ecosystem, and to patchy habitats in human-occupied landscapes. Now conjure up processes that might link cacti with cats, adding in dead prey, bacteria, and soil before working outward to water and sunlight, to phosphorus and other nutrients. As Peter Raven underscores in his essay, there's plenty going on out there and a lot about which we should be worried.

Humans have in fact been intimately part of that interconnected organic wholeness for more than two million years in Africa and Eurasia, fifty thousand in Australia, and fifteen thousand in North and South

America. Anthropogenic fires have shaped forests and grasslands over millennia, and no doubt we always have interacted in countless ways with other species. Some palms prevail in Amazonia, for example, because ten thousand years ago people planted them for food and shelter. California island foxes diverged from mainland relatives seven thousand years ago, thanks to over-water transport by Chumash, who perhaps used the little canids as granary mousers and their pelts in funerary rites. And although Lucy is our most celebrated fossil relative—she proves we first became upright, later evolved big brains—one of her closest kin reminds us that dying has always been part of living. Punctures in eye sockets of the three-million-year-old Taung child precisely match talon damage on modern monkey skulls found in eagle nests.

It also happens, owing to cultural traditions and individual backgrounds, that we favor certain landscapes thanks to their endemic critters, trees, and so forth—as conservation psychologist Susan Clayton explains in her essay, we take on particular *environmental identities*. I first met western diamond-backed rattlesnakes and gray foxes as a Hill Country grade schooler and was instantly hooked on natural born killers and open locales. As an adult I've studied in, taught about, been healed by, and advocated for wild places for more than half a century. Along my way maned wolves with reddish flanks and black leotard legs loped across Brazilian savannas. Golden Cape cobras longer than I am tall prowled ochre Kalahari sands. Glorious sunsets were savored with students, friends, and loved ones on four continents, yet here I am, back in my Texas homeland.

Austin, where I shop, has more than two million people, whereas Mason County, a hundred miles to the northwest with about the same square mileage, has four thousand citizens and no stop lights. Here on the Great Plains' southern edge, my neighbors and I restore over-grazed woodland-savannas with native grasses and prairie dogs. Amid the more numerous polled black cattle, some of us run multicolored longhorns who browse prickly pear fruit during droughts and defend their calves against all comers. We don't worm livestock with a poison that kills dung beetles, so cow pies disappear in minutes not months, less methane is released into the atmosphere, and nutrients are returned to soils. In this spirit of ecosystem health and coexistence, my neighbors and I hunt deer and wild pigs for food, but counter to deep-seated local norms, we treasure our rattlesnakes as grumpy fellow residents, and yearn for bears, mountain lions, and jaguars to make comebacks.

Out in the Hill Country, as afternoon summer temperatures peak, we cherish shade beneath majestic oaks. After darkness falls, we howl along with coyotes under orange harvest moons. And some day my remains will nourish this ground, neither injected with chemicals nor reduced to ashes and fumes. But here's the rub: my nature-loving neighbors and I have killed about four hundred raccoons. Our goal, rather than eliminate that species, is to encourage overall biodiversity—from plants and insects to large herbivores and top predators. I meet each condemned animal's gaze with a wrenching sense of duty, squeeze off a pistol round, then leave its carcass for vultures, burying beetles, and other recyclers. I am mindfully participating in this place and my conscience feels about as clean as a bug-splattered windshield.

———

Multiple players deserve to be heard in our contentiously hopeful eco-system, starting with the terrified raccoons who surely would prefer to survive. Shouldn't we as well, though, give voice to the myriad toads, turtles, lizards, and ground-nesting birds whose populations rebound when fewer small predators eat them? Can't we listen to jaguars, whose mid-twentieth-century extinction in Texas helped pave the way for a takeover by the masked bandits? As it happens, our last tiger cat, as ranchers called North America's largest felid, was "fat as butter" with remains of a raccoon in its stomach.

With a nod to Aldo Leopold's land ethic, which blends moral respon-sibility with relationships between people and nature, let's hear too from the rural families so often marginalized by environmentalists. These are folks who as kids sold raccoon pelts for $35 a pop to fund their fledgling social lives, back when bobwhite quail and lizards we called horny toads were abundant. These are folks whose descendants might yet farm and ranch in ways that favor biodiversity. Let's attend as well to

Indigenous peoples' relationships with nature, discussed herein by Potawatomi environmental justice scholar Kyle White, and to their inventions for surviving harsh climates, a point to which I'll return shortly. The rest of us might then, as a sort of reparation, reward them for trapping furbearers and saving wildlands. We too could then wear clothing more effective than synthetics against frigid winds, made from biodegradable natural fibers rather than shedding the microplastics that everywhere clog our planetary veins and arteries.

Still, confronting negative ecosystem effects from a population explosion of adorable carnivores requires navigating a pot-holed ethical labyrinth. One way out comes from those who believe killing sentient beings is morally wrong, full stop—according to them, people should never eat other animals or wear their skins, and let nature fix herself. Mind you, my own parka's ruff came from a raccoon found dead beside a country lane (fig. 25), yet for some vegans no use of another creature is ever acceptable. Pleadings on behalf of prey for relief from super-abundant predators likely wouldn't change their opinions, no matter the beleaguered hoppers and crawlers whom raccoons eat likely also have rich inner lives.

Another stance concedes that stark dichotomies like vegetarian versus meat eater, utilitarian versus intrinsic values, and individual lives versus species extinction aren't up to solving ecosystem-size problems. Down this second path lie disturbingly murky options. Leopold, for example, is more complicated than some who fondly quote him might expect—the author of *A Sand County Almanac* famously regretted shooting a fiery-eyed wolf but continued to hunt deer and wild turkeys with gusto. In northern Mexico's Sierra Tarahumara, Enrique Salmón's eco-kincentric Rarámuri utilize animals for labor, food, and protection from the elements. Leopold wrote of land "as a community in which we belong," one we "use with love and respect." Salmón's *Eating the Landscape* lyrically melds Indigenous spiritual attitudes with actions, gratitude with reciprocity, as does Potawatomi plant ecologist Robin Wall Kimmerer's widely acclaimed *Braiding Sweetgrass*.

Both facts and feelings are routinely mixed to support fiercely contested viewpoints like those with which we are grappling here. One has only to recall abortion, and because lives are at stake, both furbearers

FIGURE 25. Sunburst-style parka ruff, from the pelt of a road-killed New York raccoon. Photo by the author.

and their prey deserve a fair day in court. I admire raccoons for their paunchy ambling gaits, for their primate-like manual dexterity and mother-infant shenanigans, so I'll first stipulate that they are intelligent, intentional, and utterly lovable. Moreover, our raccoon crisis arose in a cultural context, such that their overabundance is not the fault of those individuals we trap. By the 1950s, thanks to relentless persecution by descendants of European colonists, Texas had lost its big cats, bears, and wolves that helped regulate populations of smaller predators. A few decades later pelt sales bottomed out, reportedly due to anti-fur activism. At the millennium only coyotes, bobcats, traffic, starvation, and disease remained as enemies of raccoons, whose diets were augmented by protein supplements that landowners put out for deer.

And to further blur cause, effect, and blame, over the past half century invasive fire ants provoked widespread loss of small native vertebrates across the southern US—an eco-catastrophe likely enhanced by growing numbers of waddling, ring-tailed carnivores.

———

On the other side of the ledger, humans have worn carnivore skins for several hundred thousand years. Indigenous people still harvest pelts from sea otters and other species for clothing. Laboratory experiments confirm fur's superiority for sunburst-style parkas—wolverines and wolves provide the warmest ruffs, but even raccoon beats the fake stuff. Indeed, my roadkill-trimmed jacket was a godsend while I was marooned for days in the ranch cabin during a 2021 winter storm that killed hundreds of Texans. And to further confound ethics with an ironic twist, bird-eating domestic cats garner plenty of publicity nowadays but, I suspect thanks to coyotes, they likely are insignificant in many rural areas. To the contrary, "yard" or "trash pandas," as suburbanites call raccoons, are scavengers—but out in the boonies they use perceptive eyes and noses to locate buried eggs, their sensitive forefeet to search out and manipulate secretive prey.

Truth be told, no other North American carnivore is so well adapted for exploiting—and over-exploiting—the breeding activities of other animals.

More than 90 percent of turtle nests along some beaches and estuaries on the Eastern Seaboard are destroyed by raccoons. From Illinois to Mexico, dozens of dead toads have been found around a single breeding pond, their bodies skillfully flipped, unzipped, and eviscerated so as to avoid poisonous back skin. Tracks with uncannily finger-like toe marks carpet these slaughter sites, as if the muddy banks had been visited by throngs of crawling children with two-inch hands. And to throw in still more ecological complexity, raccoon predation in southeastern US wetlands is so omnipresent that turtles, snakes, and birds place their nests in or near those of alligators—whereupon the massive reptiles, while protecting their own clutches, ambush smaller animals attracted to the

smorgasbord of egg sizes, shapes, and colors. If there were no top predators, like alligators, who would guard all those nests?

One of my Hill Country neighbors noticed some native vertebrates
remained rare after fire ants declined, whereas raccoons were ever more
frequently encountered and the most common roadkills. He knew of
their predatory expertise, set out to test a hunch, and shot about one for
every football field–size chunk of habitat on his six hundred acres. Over
the following two years he saw many more amphibians, reptiles, and
ground-nesting birds. His wife cheered as Texas horned lizards, believed extinct locally, showed up in the yard again. Woodhouse's toads,
also undetected for about a decade, became so abundant that, knowing
they eat insects, he stacked stones around their vegetable gardens into
quickly occupied toad condos. And although bobwhites are scarce in
many places, my neighbors and I are again flushing coveys and enjoying
the birds' piercing onomatopoeic calls.

For those still opposed to wearing fur, I too worry about creating a
market that might encourage factory farming. But what if wild pelts
were obtained so as to minimize cruelty, as with Colorado's ban on leg-
hold traps, or sourced from roadkill? What if taxes on fur sales funded
Indigenous health services and carnivore conservation? Would wolf or
wolverine be acceptable from areas where those species are thriving? As
for using a fur trade to manage predators, critics will argue the case
against raccoons isn't proven, so we must seek more so-called "compas-
sionate" solutions—but ought skeptics not shoulder some responsibility for all those doomed horned lizards and quail?

———

There are bigger philosophical fish to fry here. *Part* and *participate* come
from the Latin *pars*, or "part," and *partire*, "to divide or share," but can
we in ways that are both moral and mortal? Do sanctions against utilizing other species amount to *human exceptionalism*, the worldview that
we are superior to and thus apart from other life-forms? Does confining
words like *domesticate*, *slaughter*, and *culture* to our own species do so
too? As Jonathan Losos recounts in his essay, some biologists would

deny conservation status to Australian dingoes because humans shaped
the evolution of those wild canids, but how then should we treat California island foxes, given their evolutionary origin and cultural connections to the Chumash? Recalling this volume's over-arching themes, is
defining us out of wildness and wilderness not exceptionalism writ exceptionally large?

Like the authors assembled herein, even distinguished ethicists do not
agree on answers to these sorts of questions. Holmes Rolston, for example, winner of a Templeton Prize for reconciling science with religion, has
asserted that we owe neither wild nor domestic animals any more kindness than they'd receive from other species—a claim that, having watched
countless predation events, I find astonishing. At another extreme, Oxford professor Jeff McMahan has argued that eliminating all predators
would be a good thing because of the suffering they cause. And for a welcome third approach, Ben Minteer, our environmental philosopher coeditor, champions pragmatism and pluralism, in which consensus evolves
out of dialogues among diverse constituencies. Not for nothing my grad
school pal Hal Herzog, whose essay here addresses "the problem" with
domestic cats, titled his widely acclaimed book *Some We Love, Some We
Hate, Some We Eat: Why It's So Hard to Think Straight About Animals*.

Things seemed simpler when I learned to drive in the late 1950s, a few
years before contemplating the likes of epistemology and metaphysics
in a college philosophy course. Beetles, butterflies, and grasshoppers
were abundant, and windshields never long free of squashed bodies, a
morbid indicator of healthier ecosystems. Conservation, aside from
regulating hunting and fishing, focused on parks and preserves—a strategy now labeled *land sparing*. Sadly, today it's easier to see oncoming
traffic due to global insect declines but difficult to contemplate a widespread future for wildness without active coexistence, or *land sharing*.
Consistent with the first option, California recently outlawed fur sales, a
policy I'd once have supported. Lately though, working land potentially
as biodiverse as nearby protected areas of similar size, I favor increased
emphasis on the second strategy.

Given another lifetime, this nature lover would wear fur gotten with
minimal cruelty, in ways that benefit local people, especially Native

Americans. Stores would favor clothing inspired by animal adaptations to extreme climates, rather than derived from petroleum. Taking a cue from Marty Crump's account of her granddaughter examining a starfish, I envision state parks educating school kids about our relationships with sea otters. Later, home from their field trips, wide-eyed youngsters would lie on living room rugs—"Look Mom, I'm floating on my back in the waves!"—and mime how the marine mammals smash hard-shelled prey with stone tools. Showing off to older siblings with big words like ecosystem, they'd describe how otters regulate populations of sea urchin prey, who'd otherwise overgraze offshore kelp forests. And many years later they'd speak in reverent tones to their own children, of having once caressed an Indigenous-style parka ruff to appreciate how fur can protect both carnivores and people in frigid temperatures.

Today landscapes everywhere are threatened by human domination. Large species often disappear first. Many people no longer hunt for food and trap for clothing, nor grow their own fruits and vegetables, and idealizing those activities won't guarantee coexistence. My bet, though, is that even city folks can live more thoughtfully part of nature. Let's expand our Capital's quirky motto to "Keep Austin weird *and wild,*" reflecting its celebrated Congress Avenue bridge bats and University of Texas campus peregrines. After all, as Emma Marris observes in her essay, with luck there might go a badger. And out beyond urban areas, given that other lifetime, I'd replace the juggernaut of industrial beef with herds of bison and heritage-breed cattle (fig. 26). Given space and freedom to roam, these iconic mammals would enjoy species-typical social behaviors and employ horns to safeguard offspring from re-colonizing big cats, bears, and wolves. The top carnivores in turn would help control populations of smaller species.

Humans are *wildland gardeners* in my futuristic daydreams, to use tropical ecologist Daniel Janzen's apt shorthand. Woodland-savanna ecosystems would support regional food and clothing production, as well as house big predators and herbivores. Parks would insure against population declines, provide benchmarks for sparing versus sharing, and offer places for wilderness recreation and contemplation. Indigenous peoples would play increasingly prominent roles rewilding both

FIGURE 26. Longhorns on Rancho Cascabel in the Texas Hill Country. These semiarid-adapted cattle are descended from Iberian stock brought to North America about five hundred years ago. Photo by the author.

shared and spared lands. None of this would guarantee the return of jaguars to Texas, but in Mexico cattle with horns are less vulnerable to big cats than those who lack them and fewer livestock are taken where native peccaries provide abundant prey. Likewise, one of my neighbors, during a quarter century of living with longhorns, has never lost a calf to predators, but he did once find a dead coyote near his herd, gored through the chest.

In other words, pulling together ecological, cultural, economic, and political pieces of a wickedly difficult puzzle—conserving predators on working landscapes—well, maybe someday . . .

In the meantime, inspired by the Indigenous perspectives of Salmón and Kimmerer, "attitudes with actions, gratitude with reciprocity" feels to me like a worthy mantra for these perilous times. Last year delivered unprecedented heat and drought to the Hill Country, further evidence that we face daunting challenges from climate change. Month after interminable month, previously permanent ponds stayed dry and the only water for wildlife was in cattle tanks filled from ranchers' wells. Spring brought almost none of the bluebonnets and other wildflowers for which our region is famous. When summer temperatures hovered above 100°F for week after scorching week, my longhorns used their hooves to construct shallow resting pits in soil under the larger oaks, evidently aware of what biophysicists call conductive cooling. Familiar frogs finally called when fall rains made dirt roads briefly impassable.

We had a few good omens as 2022 ended, grim tidings notwithstanding. October yielded my first encounter at the ranch with an eastern hog-nosed snake, a species whose behavior I had studied in graduate school. These serpents use their spade-like snouts to dig up toads, and she looked well fed, further evidence those amphibians are once again common. Then, on a starlit December solstice, coyotes chorused a few hundred yards from my cabin—one pack to the east, another to the south, yipping and yodeling as they had almost nightly for the past year. Raccoons, on the other hand, have been less evident since the dry spell, so for now at least I'm giving them a break. With thanks to that one ill-fated animal for his pelt and the northern people who invented parka ruffs, here's hoping we all make it through another winter and beyond.

ACKNOWLEDGMENTS

A hearty thanks to our contributors for their hard work and engagement over the course of this project. We greatly appreciate the suggestions of two anonymous reviewers of the manuscript for Princeton University Press; their recommendations made this an even stronger discussion. We'd also like to thank the Press's Alison Kalett and Hallie Schaefer for their steady support and guidance, and Christie Henry for encouraging us as the idea for the volume took initial shape. Finally, we'd like to acknowledge the support of the Arizona Zoological Society Endowed Chair position at Arizona State University (Ben) and the William H. Danforth Distinguished University Professorship at Washington University in St. Louis (Jonathan).

NOTES

Introduction: "Wild Hearts and Minds," Minteer and Losos

The discussion of Cole's background and arrival in the US is taken from Angela Miller, "The Fate of Wilderness in American Landscape Art. The Dilemmas of 'Nature's Nation'," in *American Wilderness: A New History*, ed. Michael Lewis (Oxford, UK: Oxford University Press, 2007), 91–112; and Tim Barringer, "Thomas Cole's Atlantic Crossings," in *Thomas Cole's Journey. Atlantic Crossings*, ed. Elizabeth Mankin Kornhauser and Tim Barringer (Metropolitan Museum of Art / Yale University Press, 2018), 19–61. Cole's experience in the Catskills is described in Roderick Frazier Nash, *Wilderness and the American Mind*, 5th ed. (New Haven: Yale University Press, 2014), 78–80. His response to Jacksonian America is discussed in Barringer, "Thomas Cole's Atlantic Crossings." Cole's deep interest in geology and natural history is referenced in Elizabeth Mankin Kornhauser, "Manifesto for an American Sublime: Thomas Cole's *The Oxbow*" in Kornhauser and Barringer, eds., *Thomas Cole's Journey*, 63–95; and Linda S. Ferber, *The Hudson River School. Nature and the American Vision* (New York Historical Society / Rizzoli, 2009). The effect of environmental destruction in the Northeast on Cole's thinking and art is discussed in Kornhauser, "Manifesto for an American Sublime."

Cole's remark, "the most distinctive, and perhaps the most impressive, characteristic of American scenery is its wildness" appears in his 1836 essay, "Essay on American Scenery," originally published in *American Monthly* magazine. Full text is available at: https://thomascole.org /wp-content/uploads/Essay-on-American-Scenery.pdf. Rod Nash's analysis of Cole's fears about losing the connection to the wilderness, as well the tension between the wild and civilized in his thinking, appears in *Wilderness and the American Mind*, 79–82. The question mark motif in Cole's *The Oxbow* is referenced (with different interpretations) in Don Scheese, *Nature Writing: The Pastoral Impulse in America* (New York: Routledge, 2002) and Barringer, "Thomas Cole's Atlantic Crossings."

Chapter 2: "Embracing the Cane Toad," Shine

Many people have written about shifting baselines in environmental issues. Seminal articles include those by D. Pauly, for example, "Anecdotes and the Shifting Baseline Syndrome of Fisheries," *Trends in Ecology and Evolution* 10 (1995): 430; T. Hawke, G. Bino, and R. T. Kingsford, "A Silent Demise: Historical Insights into Population Changes of the Iconic Platypus (*Ornithorhynchus anatinus*)," *Global Ecology and Conservation* 20 (2019): 00720; and S. Lovell, A. E. Johnson, R. Ramdeen, and L. McClenachan, "Shifted Baselines and the Policy Placebo Effect in Conservation," *Oryx* 54 (2020): 383–391. For an examination of novel ecosystems, see R. J. Hobbs, E. S. Higgs, and C. Hall, *Novel Ecosystems: Intervening in the New Ecological World Order* (Hoboken, NJ: John Wiley & Sons, 2013). An anthology edited by Gavin Van Horn and John Hausdoerffer (*Wildness*, University of Chicago Press, 2017) explores the idea that we may need to jettison our existing views about wilderness to cope with a changing planet and society.

For a review of the idea that our brains compare reality to expectation, see A. Clark, *Surfing Uncertainty: Prediction, Action, and the Embodied Mind* (New York: Oxford University Press, 2015).

For discussions about how children perceive the natural world, take a look at J-M. Ballouard, F. Brischoux, and X. Bonnet, (2011). "Children Prioritize Virtual Exotic Biodiversity over Local Biodiversity," *PLoS One* 6, no. 8 (2011): e23152, doi:10.1371 / journal.pone.0023152; and M. Genovart, G. Tavecchia, J. J. Ensenat, and P. Laiolo, "Holding Up a Mirror to the Society: Children Recognize Exotic Species Much More Than Local Ones," *Biological Conservation* 159 (2013): 484–89, doi:10.1016/j.biocon.2012.10.028.

And for papers about my own research projects that found their way into this essay, here are some of the relevant ones:

G. P. Brown, and R. Shine, "Repeatability and Heritability of Reproductive Traits in Free-Ranging Snakes," *Journal of Evolutionary Biology* 20 (2007): 588–96.

R. Shine, "Sex at the Snake Den: Lust, Deception and Conflict in the Mating System of Red-Sided Garter Snakes," *Advances in the Study of Behaviour* 44 (2012): 1–51.

R. Shine, and L. Sun, "Arboreal Ambush Site Selection by Pit-Vipers (*Gloydius shedaoensis*)," *Animal Behaviour* 63 (2002): 565–76.

R. Shine, L. Sun, E. Zhao, and X. Bonnet, "A Review of 30 Years of Ecological Research on the Shedao Pit-Viper," *Herpetological Natural History* 9 (2002): 1–14.

R. Shine, T. G. Shine, G. P. Brown, and C. Goiran, "Life History Traits of the Sea Snake *Emydocephalus annulatus*, Based on a 17-Year Study," *Coral Reefs* 39 (2020): 1407–14.

Chapter 3: "Invasive Species in the Anthropocene, or Learning to Love the Dingo," Losos

Geerat Vermeij reviewed biological invasions in "When Biotas Meet: Understanding Biotic Interchange," *Science* 253 (1991): 1099–1104. New views on introduced species and biodiversity conservation can be found in Davis et al., "Don't Judge Species on Their Origins," *Nature* 474 (2011): 153–54; F. Pearce, *The New Wild: Why Invasive Species Will Be Nature's Salvation* (London: Icon Books, 2015); and C. D. Thomas, *Inheritors of the Earth: How Nature Is Thriving in an Age of Extinction* (Westminster: Penguin Books, 2017). A recent critique is C. H. Lean, "Invasive Species Increase Biodiversity and, Therefore, Services: An Argument of Equivocations," *Conservation Science and Practice* 3 (2021): e553. The quote from Dan Simberloff is in "Nature, Culture, and Natureculture: The Role of Nonnative Species in Biocultures," in *From Biocultural Homogenization to Biocultural Conservation*, ed. R. Rozzi et al. (Cham, Switzerland: Springer Nature), 207–18. David Quammen wrote a prescient article on biodiversity travails in which he coined the term "Planet of Weeds" (*Harper's*, October 1998). A nice introduction to the argument that dingoes belong in Australia is Peter Banks, "An Eco-Evolutionary Rationale to Distinguish Alien and

Native Status: Why the Dingo Is a Native Species on Mainland Australia," *Australian Zoologist* 41 (2021): 358–66. A review of the history of giant tortoise evolution and their use in ecosystem restoration is Hansen et al., "Ecological History and Latent Conservation Potential: Large and Giant Tortoises as a Model for Taxon Substitutions," *Ecography* 33 (2010): 272–84.

Chapter 4: "Bringing the Wild Things into Our Lives: The Problem with Cats," Herzog

Research on domestic cats and our relationships with them has exploded in recent years. For overviews, see *Cat Sense: How the New Feline Science Can Make You a Better Friend to Your Cat*, by John Bradshaw (New York: Basic Books, 2013), and *The Domestic Cat: The Biology of Its Behavior*, edited by Dennis Turner and Patrick Bateson (Cambridge: Cambridge University Press, 2014). Jessica Pierce's book *Run, Spot, Run: The Ethics of Keeping Pets* (Chicago: University of Chicago Press, 2016) and *Companion Animal Ethics*, edited by Peter Sandøe, Sandra Corr, and Clare Palmer (Hoboken, NJ: John Wiley & Sons, 2015) are excellent sources of information on ethical issues associated with bringing cats and other companion animals into our lives. For information on the changing roles of cats in Western culture, see *Cats*, by Katherine M. Rogers (London: Reaktion Press, 2006). *Cat Wars: The Devastating Consequences of a Cuddly Killer*, by Peter Marra and Chris Santella (Princeton: Princeton University Press, 2016) is the most influential treatment of the impact of both pet and feral cats on wildlife.

The introductory quote, "Something from every cat still vibrates with its wildness," is from Peter Christie's book *Unnatural Companions: Rethinking Our Love of Pets in an Age of Wildlife Extinction* (Washington, DC: Island Press, 2020.) The discussion of the evolution of domestic cats is based on C. A. Driscoll, D. W. Macdonald, and S. J. O'Brien, "From Wild Animals to Domestic Pets, an Evolutionary View of Domestication," *Proceedings of the National Academy of Sciences* 106, Suppl. 1 (2009), 9971–78. The claim that cats are semi-domesticated is found in J. A. Serpell, "Domestication and History of the Cat," in *The Domestic Cat: The Biology of Its Behaviour*, ed. D. C Turner and P. Bateson (Cambridge, UK: Cambridge University Press), 83–100. Information on the treatment of cats from ancient Egypt through the nineteenth century is based on J. Bradshaw, *Cat Sense: How the New Feline Science Can Make You a Better Friend to Your Cat* (New York: Basic Books, 2013). The health problems of flat-faced cat breeds are described in D. G. O'Neill, C. Romans, D. C. Brodbelt, D. B. Church, P. Černá, and D. A. Gunn-Moore, "Persian Cats under First Opinion Veterinary Care in the UK: Demography, Mortality, and Disorders," *Scientific Reports* 9 no. 1 (2019): 1–13, and in L. Plitman, P. Černá, M. J. Farnworth, R. Packer, and D. A. Gunn-Moore, (2019). "Motivation of Owners to Purchase Pedigree Cats, with Specific Focus on the Acquisition of Brachycephalic Cats," *Animals* 9, no. 7 (2019): 394. The development and popularity of wild cat X domestic cat hybrids is described by Ariel Levy in "Living-Room Leopards" in the *New Yorker* (April 29, 2013). The potential impact of Savannah cats on Australian wildlife is documented in C. R. Dickman, S. M. Legge, and J. C. Woinarski, "Assessing Risks to Wildlife from Free-Roaming Hybrid Cats: The Proposed Introduction of Pet Savannah Cats to Australia as a Case Study," *Animals* 9, no. 10 (2019): 795. Estimates of the total number of cats on Earth vary from 500 million to a billion. The 370 million pet cats figure is based on research by Andrew Rowan (personal communication, June, 2021). The anthropological study of the roles of pets in sixty cultures is by P. B. Gray and S. M. Young, "Human-Pet Dynamics in Cross-Cultural Perspective," *Anthrozoös* 24, no. 1 (2011): 17–30.

The now-classic paper on the tolls cats have on birds and mammals is by S. R. Loss, T. Will, and P. P. Marra, "The Impact of Free-Ranging Domestic Cats on Wildlife of the United States," *Nature Communications* 4, no. 1 (2013): 1–8. The case that impact of cats on wildlife is overblown is made by W. S. Lynn, F. Santiago-Ávila, J. Lindenmayer, J. Hadidian, A. Wallach, and B. J. King, "A Moral Panic over Cats," *Conservation Biology* 33, no. 4 (2019): 769–76. The authors' response to their critics is in S. R. Loss, T. Will, T. Longcore, and P. P. Marra, "Responding to Misinformation and Criticisms Regarding United States Cat Predation Estimates," *Biological Invasions* 20, no. 12 (2018): 3385–96. For research showing most pet cats are not interested in or good at hunting, see K.A.T. Loyd, S. M. Hernandez, J. P., Carroll, K. J., Abernathy, and G. J. Marshall, "Quantifying Free-Roaming Domestic Cat Predation Using Animal-Borne Video Cameras," *Biological Conservation* 160 (2013): 183–89. For an overview of the health and behavioral impact of keeping cats indoors 24 / 7, see S. M. Tan, A. C. Stellato, and L. Niel, "Uncontrolled Outdoor Access for Cats: An Assessment of Risks and Benefits," *Animals* 10, no. 2 (2020): 258, and P. Sandøe, A. P. Nørspang, B. Forkman, C. R. Bjørnvad, S. V. Kondrup, and T. B. Lund, "The Burden of Domestication: A Representative Study of Welfare in Privately Owned Cats in Denmark," *Animal Welfare* 26, no. 1 (2017): 1–10. The finding that indoor cats are prone to repetitive behavior disorders is found in L. R. Kogan, E. K. Grigg, "Laser Light Pointers for Use in Companion Cat Play: Association with Guardian-Reported Abnormal Repetitive Behaviors," *Animals* 11, no. 8 (2021): 2178. The national comparisons of the percentage of pet cats allowed outdoors are from A. N. Rowan, T. Kartal, and J. Hadidian, "Cat Demographics and Impact on Wildlife in the USA, the UK, Australia and New Zealand: Facts and Values," *Journal of Applied Animal Ethics Research* 2, no. 1 (2019): 7–37. Bernard Rollin's statement, "*Fish gotta swim, birds gotta fly,*" is from B. E. Rollin, "The Moral Status of Invasive Animal Research," *Hastings Center Report* 42, s1 (2012): S4–S6. The ethicist David Degrazia's argument against keeping cats indoors is found in D. DeGrazia, (2011) "The Ethics of Confining Animals: From Farms to Zoos to Human Homes," in *The Oxford Handbook of Ethics and Animals*, ed. Tom L. Beauchamp and R. G. Frey (New York: Oxford University Press), 738–68. C. E. Abbate's perspective on why cat owners have an obligation to allow their pet to roam outdoors is in C. E. Abbate, "A Defense of Free-Roaming Cats from a Hedonist Account of Feline Well-Being," *Acta Analytica* 35, no. 3 (2020): 439–461. Information on breed differences in the predatory behavior in cats is found in B. L. Hart and L. A. Hart, *Your Ideal Cat: Insights into Breed and Gender Differences in Cat Behavior* (West Lafayette, IN: Purdue University Press, 2013); and D. L. Duffy, R.T.D. de Moura, and J. A. Serpell, "Development and Evaluation of the Fe-BARQ: A New Survey Instrument for Measuring Behavior in Domestic Cats (*Felis s. catus*)," *Behavioural Processes* 141 (2017): 329–41. Research on the role of genetics in individual differences in cats is from M. Salonen, K. Vapalahti, K. Tiira, A. Mäki-Tanila, and H. Lohi, "Breed Differences of Heritable Behaviour Traits in Cats," *Scientific Reports* 9, no. 1 (2019): 1–10. The experiment showing that playing with cats and giving them high-meat diets reduces predation is in M. Cecchetti, S. L. Crowley, C. E. Goodwin, and R. A. McDonald, "Provision of High Meat Content Food and Object Play Reduce Predation of Wild Animals by Domestic Cats *Felis catus*," *Current Biology* 31, no. 5 (2021): 1107–11. The case that the relationship between cats and wildlife is a wicked problem is found in W. S. Lynn and F. J. Santiago-Ávila, "Outdoor Cats: Science, Ethics, and Politics," *Society and*

Animals 30, no. 7 (2022): 798–815. The New Zealand Cook's petrel case is discussed in M. J. Rayner, M. E. Hauber, M. J. Imber, R. K. Stamp, and M. N. Clout, "Spatial Heterogeneity of Mesopredator Release within an Oceanic Island System," *Proceedings of the National Academy of Sciences* 104, no. 52 (2007): 20862–65.

Chapter 5: "How Did We Get Here?," Raven

There are dozens of excellent works on these subjects that I have consulted during the course of my career. Recent ones that I have found to be of particular interest include Enrique Sala's *The Nature of Nature. Why We Need the Wild* (Washington, DC: National Geographic Society, 2020), which presents as clear a picture of the relationships between species and the ecosystems they compose as any work I have read. I also found Tony Hiss's *Rescuing the Planet. Protecting Half the Land to Heal the Earth* (New York: Alfred A. Knopf, 2021) provided for me an inspiration with numerous fine examples of impressive efforts to organize and preserve significant stretches of land, including some around cities, to balance nature with humanity. Pope Francis's encyclical "Laudato Si" presents a moving view of the importance of saving the earth from a religious point of view, and is well worth reading. Patriarch Bartholomew I of the Eastern Orthodox Church has been particularly active, and both he and Pope Francis have declared destroying the environment a sin (https://www.washingtonpost.com/national/health-science/climate-change-is-a-top-spiritual-priority-for-these-religious-leaders/2018/06/26/d5e06fd2-749e-11e8-9780-b1dd6a09b549_story.html?utm_term=.9fbb6445d473). For them, as for many others, global warming, which promises to alter drastically the conditions and challenges faced by all life in the future, has provided a strong stimulus to action. My recently published autobiography, *Driven by Nature* (St. Louis: Missouri Botanical Garden Press, 2021), traces the evolution of ideas in this field over the past century, and offers suggestions for the future.

Chapter 6: "Why Does Anything Need to be Called Wild?," Whyte

Dr. Bernard Perley retains the copyright to the cartoon referenced on the first page of this essay, which refers to the 2019 version of the cartoon adapted from Perley's earlier versions. In an email, Perley granted permission to the author (Whyte) to reference the cartoon only in this essay. The author and publisher do not have permission to republish the cartoon in any other form beyond its reference in this essay. Other references include Basil Johnston's *Honour Earth Mother* (Lincoln: University of Nebraska Press, 2003); Anya Zilberstien's "Inured to Empire: Wild Rice and Climate Change," *William and Mary Quarterly* 72, no. 1 (2015): 127–58; Mike Dockry and Kyle Whyte's "Improving on Nature," *American Indian Quarterly* 45, no. 2 (2021): 95–120; Eric Freedman's "When Indigenous Rights and Wilderness Collide: Prosecution of Native Americans for Using Motors in Minnesota's Boundary Waters Canoe Wilderness Area," *American Indian Quarterly* 26, no. 3 (2002): 378–92; Leanne Simpson's "Looking after Gdoo-Naaganinaa," in *Wicazo Sa Review* 3, no. 2 (2008): 29–42; Heidi Stark, "Respect, Responsibility, and Renewal," *American Indian Culture and Research Journal* 34, no. 2 (2010): 145–64; Michael Johnson, "Fire over Ahwahnee," *Scientific American*, August 13, 2014, https://blogs.scientificamerican.com/primate-diaries/how-john-muir-s-brand-of-conservation-led-to-the-decline-of-yosemite/; Megan Bang et al., "Muskrat Theories, Tobacco in the Streets, and Living Chicago as Indigenous Land," *Environmental Education Research* 20, no. 1 (2014): 37–55; Clint Carroll, "Native Enclosures: Tribal National Parks and the Progressive Politics of Environmental Stewardship in Indian Country," *Geoforum* 53 (2014): 31–40.

Chapter 7: "Affirming the Wilderness Ideal," Crist

Influential wilderness critique papers include Ramachandra Guha's "Radical American Environmentalism and Wilderness Preservation: A Third World Critique," *Environmental Ethics* 11 (1989): 71–83; William Denevan's "The Pristine Myth: The Landscape of the Americas in 1492," *Annals of the Association of American Geographers* 82, no. 3 (1992): 369–85; and, perhaps most famously, William Cronon's "The Trouble with Wilderness; or, Getting Back to the Wrong Nature," in *Uncommon Ground: Rethinking the Human Place in Nature*, ed. W. Cronon (New York: W. W. Norton, 1995), 69–90. For an accessible pro-wilderness presentation of "biological wilderness," and a clear comparative discussion of wilderness, wildness, and biodiversity, see Reed Noss's lecture "Wilderness, Wildness, and Biodiversity: We Need All Three," presented by the University of Montana Wilderness Institute, 2013, https://www.youtube.com/watch?v=UcCUX_w5uhw. For a more popular defense of wilderness, and a plea not to abandon the idea, see science reporter Brandon Keim's "Earth Is Not a Garden," *Aeon*, September 18, 2014, https://aeon.co/essays/giving-up-on-wilderness -means-a-barren-future-for-the-earth. Gary Snyder's *The Practice of the Wild* (New York: North Point Press, 1990); and Jack Turner's *The Abstract Wild* (Tucson: University of Arizona Press, 1996) offer poetic and comprehensive defenses of wilderness and wildness. For strong philosophical critiques of the Western idea that nonhuman nature is inherently "meaningless," see Erazim Kohák's *The Ember and the Stars: A Philosophical Inquiry into the Moral Sense of Nature* (Chicago: University of Chicago Press, 1987) and Freya Mathews's *Reinhabiting Reality: Towards a Recovery of Culture* (Albany: State University of New York Press, 2005). The paradigm of "new environmentalism" was arguably first laid out in a widely read paper by Peter Kareiva, Robert Lalasz, and Michelle Marvier titled "Conservation in the Anthropocene: Beyond Solitude and Fragility," *Breakthrough Journal* (Fall 2011): 29–37. Much of the Anthropocene literature in the past decade has emphasized portraying humanity as a "force of nature," and even calling humans "god-like" as Mark Lynas (among others) does in his book, *The God Species: How the Planet Can Survive the Age of Humans* (London: Fourth Estate, 2011). In his human ecology work, geographer Erle Ellis stresses the point that humanity has had a long and formative impact on the planet. A solid scientific treatment of this assessment can be found in Ellis's coauthored paper "Used Planet: A Global History," *PNAS* 110, no. 20 (2012): 7978–85. A popular exposition of the same idea can be found in Ellis's "The Long Anthropocene: Three Millennia of Humans Reshaping the Earth," 2013, http://thebreakthrough.org/index .php/programs/conservation-and-development/the-long-anthropocene. Mark Tercek's article, "Money Talks—So Let's Give Nature a Voice," is published in *Earth Island Journal* (Summer 2013), https://www.earthisland.org/journal/index.php/magazine/entry/money_talks_so_lets_give _nature_a_voice/. Peter Kareiva and Michelle Marvier's plea for "realism"—essentially for resigning ourselves to a biodiversity-impoverished world—is cited from their paper "What Is Conservation Science?" *Bioscience* 62, no. 11 (2011): 962–69. In an overview paper, titled "The Battle for the Soul of Conservation Science," *Issues in Science and Technology* 31, no. 2 (Winter 2015): 74–79, Keith Kloor discusses the contemporary debate between the viewpoints of protecting biodiversity for its own sake versus for the sake of human material well-being. Peter Kareiva's statement that mass extinctions eventually lead to "a new evolutionary burst" is quoted in Kloor's article. The similar claim by scientist Chris Thomas is reproduced from his interview with Fred Pearce, "Human Meddling Will Spur the Evolution of New Species," *New Scientist*, January 8, 2014, https://www .newscientist.com/article/mg22129510-400-human-meddling-will-spur-the-evolution-of-new -species/. Kareiva and Thomas neglect to mention that the resurgence of another biodiversity chapter after a mass extinction event takes millions of years—a timeline essentially irrelevant to all future human generations. A positive spin on the Anthropocene, as a time rife with human opportunity, is offered by Ellis in his "The Planet of No Return: Human Resilience on an Artificial Earth," *The Breakthrough Institute*, January 6, 2012, https://thebreakthrough.org/journal/issue-2/the-planet-of -no-return, and by reporter Andrew Revkin (in his plea for a "good Anthropocene") in a *New York Times* editorial "Confronting the 'Anthropocene'" (May 11, 2011). Calls for optimism and hope in our time can certainly be heartening, but they should not come at the expense of understating the gravity of the nonhuman and human predicament today. Michael Shellenberger and Ted Nordhaus's entreaty to regard technology as "sacred" is articulated in their ebook *Love Your Monsters: Postenvironmentalism and the Anthropocene*. Ellis's quotes endorsing technology as human destiny are from his article "Overpopulation Is Not the Problem," *New York Times*, September 13, 2013. Soulé's protest against smearing pro-wilderness environmentalism as "a dysfunctional antihuman anachronism" is cited in an article by D. T. Max ("Green Is Good," *The New Yorker*, May 5, 2014). The paraphrase of John Rodman is from wording in his paper, "The Liberation of Nature?," *Inquiry* 20 (1987): 83–145. For broad and accessible perspectives on rewilding, see Caroline Fraser's book *Rewilding the World: Dispatches from the Conservation Revolution* (New York: Picador, 2009); George Monbiot's book *Feral: Searching for Enchantment on the Frontiers of Rewilding* (London: Allen Lane, 2013); and David Johns, "History of Rewilding: Ideas and Practice," in *Rewilding*, ed. N. Pettorelli, S. Durant, and J. Du Toit, 12–33 (Cambridge: Cambridge University Press, 2019), https://www .cambridge.org/core/books/abs/rewilding/history-of-rewilding-ideas-and-practice/495C70887 21B4390D37504BBF24EB01D.

For a comprehensive scientific overview of the growth of human systems, and related adverse human impacts on the Earth system, see Will Steffen and colleagues, "The Trajectory of the Anthropocene: The Great Acceleration," *Anthropocene Review* 2, no. 1 (2015): 81–98. While new environmentalists tend to accept the expansionist trends of the "Great Acceleration," and argue for managerial and technological fixes to problems, rewilding advocates contend that we need to focus primarily on downscaling and radically transforming the human enterprise in order to heal humanity's relationship with the Earth.

Chapter 8: "Picturing the Wild," Minteer

Biographical material on Adams's childhood and first visit to Yosemite was drawn from Ansel Adams (with Mary Street Alinder), *An Autobiography* (New York: Little, Brown, 1996); Mary Street Alinder, *Ansel Adams* (New York: Bloomsbury, 2014); and Jonathan Spaulding, *Ansel Adams and the American Landscape* (Berkeley: University of California Press, 1995). The Adams quote in the figure 11 caption appears in *An Autobiography*, 122. The quote in the figure 12 caption appears in "Give Nature Time," his 1961 commencement address to Occidental College, reprinted in *Ansel*

Adams and the National Parks, ed. Andrea G. Stillman (New York: Little, Brown, 2010), 12. The quote in the caption for figure 13 comes from his essay "The Artist and the Ideals of Wilderness," in *Wilderness. America's Living Heritage,* ed. David Brower (San Francisco: Sierra Club, 1961), 13.

Adams's photographic style, including in his photographs of the national parks in the 1940s, is discussed in Rebecca Senf's excellent study, *Making a Photographer: The Early Work of Ansel Adams* (New Haven: Yale University Press, 2020); and Anne Hammond, *Ansel Adams: Divine Performance* (New Haven: Yale University Press, 2002). The Adams passage, "Photography, in conjunction with the appropriate written word and suitable media for distribution," is from "The Artist and Ideals of Wilderness" (49). William Turnage's remarks about *This Is the American Earth* appear in his essay, "Ansel Adams, Environmentalist" in *Ansel Adams in the National Parks.* Finis Dunaway's discussion of the Sierra Club Exhibit Format series and its role in environmental advocacy appears in his fine work *Natural Visions. The Power of Images in American Environmental Reform* (Chicago: University of Chicago Press, 2005). J. B. Jackson's critique of those same books is found in his 1995 collection *A Sense of Place, a Sense of Time* (New Haven: Yale University Press, 1994).

William Cronon's "The Trouble with Wilderness; or, Getting Back to the Wrong Nature" was published in *Uncommon Ground: Rethinking the Human Place in Nature,* ed. W. Cronon (New York: W. W. Norton, 1996). Deborah Bright's remark comes from "The Machine in the Garden Revisited: American Environmentalism and Photographic Aesthetics," *Art Journal* 51 (1992): 60–71. Mark Dowie's critique of Adams and the classical photographers of Yosemite appears in his essay "The Myth of a Wilderness without Humans," *MIT Press Reader,* 2011, https://thereader.mitpress.mit.edu/the-myth-of-a-wilderness-without-humans/.

Rebecca Solnit's observation about Adams's cropping in Yosemite can be found in Mark Klett, Rebecca Solnit, and Byron Wolfe, *Yosemite in Time* (San Antonio: Trinity University Press, 2005). Wallace Stegner's defense of Adams as a photographer appears in his foreword to *Ansel Adams: Images, 1923–74* (Boston: New York Graphic Society, 1974). Adams's remark "The appreciation of the natural scene should not be limited only to the grandeurs" is taken from his 1961 Occidental commencement address. The Diablo Canyon controversy within the Sierra Club is discussed in Michael P. Cohen, *The History of the Sierra Club, 1892–1970* (San Francisco: Sierra Club Books, 1988); Tom Turner, *David Brower: The Making of the Environmental Movement* (Berkeley: University of California Press, 2015); and Spaulding, *Ansel Adams and the American Landscape.* Adams's 1969 letter about the Diablo case is collected in *Ansel Adams, Letters 1916–1984* (Boston: Little, Brown, 1988). I'm indebted to photographer and friend Mark Klett for his personal observation that Adams taught people *how* to see wild nature, including how to admire it. The strong statement arguing for a muscularly humanistic environmental ethic in the Anthropocene is found in Erle Ellis, "Too Big for Nature" in *After Preservation: Saving American Nature in the Age of Humans,* ed. Ben A. Minteer and Stephen J. Pyne (Chicago: University of Chicago Press, 2015).

Chapter 9: "In Feral Land Is the Preservation of the World," Moore

The statistics about species loss are hard to believe, but they come from a reliable source, the 2019 United Nations report *Nature's Dangerous Decline 'Unprecedented'; Species Extinction Rates 'Accelerating.'* https://www.un.org/sustainabledevelopment/blog/2019/05/nature-decline-unprecedented-report/. Likewise, the statistics on the rate of extinction come from a solid organization, the World Wildlife Fund: https://wwf.panda.org. Eileen Crist's important book, *Abundant Earth: Toward an Ecological Civilization* (University of Chicago Press, 2019), provides detailed information about the philosophical views that underpin the climate catastrophes. The Natural History Museum of London is the source of the percentage of species obliterated during the fifth extinction events. The essential and informative website of the Half-Earth Project explains how to save 84 percent of the earth's species: https://eowilsonfoundation.org/. For information about the rough-skinned newt (*Taricha granulosa*), I am privileged to turn to my husband, Frank L. Moore, distinguished professor emeritus, Oregon State University, an international expert on the endocrine control of reproductive behavior in amphibians. Bill McKibben's book about the ubiquitous human presence is, of course, *The End of Nature* (Random House, 2006).

Chapter 10: "Revealing and Reveling in the Story of Nature," Fleischner

The Mogollon Highlands are described in T. L. Fleischner et al., "The Mogollon Highlands Ecoregion of the American Southwest: A Neglected Center of Ecological Diversity," *Natural Areas Journal* 44(2) (April 2024).

Paul Shepard discussed the evolutionary origins of human consciousness in *Thinking Animals: Animals and the Development of Human Intelligence* (New York: Viking Press, 1978; see, especially, 1–12).

Paul Shepard's phrase, "Ten Thousand Years of Crisis," is the title of the first chapter in *The Tender Carnivore and the Sacred Game* (New York: Charles Scribner's Sons, 1973).

Pliny the Elder's complete *Historia Naturalis* is available online through, among other outlets, the US National Library of Medicine (part of the National Institutes for Health), and the Natural History Museum in the UK, https://www.nhm.ac.uk/discover/news/2016/december/museums-oldest-natural-history-book-now-accessible-online.html. A single volume, *Natural History: A Selection*, is still in print from Penguin Classics, John F. Healy, translator, 1991.

I summarized many definitions of natural history in: "Natural History and the Deep Roots of Resource Management," *Natural Resources Journal* 45 (2005): 1–13. Among the many voices calling attention to the importance of natural history: Harry W. Greene, "Natural History and Evolutionary Biology," in *Predator-Prey Relationships: Perspectives and Approaches from the Study of Lower Vertebrates*, ed. M. E. Feder and G. V. Lauder (Chicago: University of Chicago Press, 1986), 99–108; and "Organisms in Nature as a Central Focus for Biology," *Trends in Ecology and Evolution* 20 (2005): 23–27; T. L. Fleischner, "Revitalizing Natural History," *Wild Earth* 9, no. 2 (Summer 1999): 81–89; and Steven G. Herman, "Wildlife Biology and Natural History: Time for a Reunion," *Journal of Wildlife Management* 66 (2002): 933–46; Joshua Tewksbury et al., "Natural History's Place in Science and Society," *BioScience* 64 (2014): 300–310; Seabird McKeon et al., "Human Dimensions: Natural History as the Innate Foundation of Ecology," *Bulletin of the Ecological Society of America* 101 (2020): 1–7.

The effect of Linnaeus's taxonomic system on exploration and discovery is discussed in Howard Ensign Evans, *Pioneer Naturalists: The Discovery and Naming of North American Plants and Animals* (New York: Henry Holt, 1993). Natural history's challenging of assumptions is explored in Ernst Mayr, *The Growth of Biological Thought* (Cambridge, MA: Belknap / Harvard University Press, 1982).

The word "scientist" was coined in 1840 by the English philosopher and mathematician William Whewell, and "suggested a growing professional consciousness." See Donald Worster, *Nature's Economy: A History of Ecological Ideas* (Cambridge, MA: Cambridge University Press, 1977), 130.

An excellent historical resource is John G. T. Anderson's *Deep Things out of Darkness: A History of Natural History* (Berkeley: University of California Press, 2013). I summarized key elements of this story in "Revitalizing Natural History," *Wild Earth* 9, no. 2 (Summer 1999): 81–89.

Melissa K. Nelson and Dan Shilling, eds., *Traditional Ecological Knowledge: Learning from Indigenous Practices for Environmental Sustainability* (Cambridge, MA: Cambridge University Press, 2021) provides a good review of the scope of this approach.

Charles Elton's *Animal Ecology* was first published in 1927 (London: Sidgwick & Jackson), 1.

Aldo Leopold's speech on natural history as a forgotten science later found its way into his book, *Round River*, ed. Luna B. Leopold (Minocqua, WI: NorthWord Press 1991), 92–101.

George Bartholomew discussed the limits of hypothesis testing in biological sciences in "The Role of Natural History in Contemporary Biology," *BioScience* 36 (1986): 324–29.

My definition of natural history and argument for its importance was first presented in "Natural History and the Spiral of Offering," *Wild Earth* 11, nos. 3–4 (2001): 10–13, then elaborated upon in "Natural History and the Deep Roots of Resource Management," *Natural Resources Journal* 45 (2005): 1–13; "The Mindfulness of Natural History," in *The Way of Natural History* (San Antonio, TX: Trinity University Press, 2011), 3–15; "Why Natural History Matters," *Journal of Natural History Education and Experience* 5 (2011): 21–24; and "Our Deepest Affinity," in *Nature, Love, Medicine: Essays on Wildness and Wellness* (Salt Lake City: Torrey House Press, 2017), 3–15.

Early US conservation policy was described by James B. Trefethen, *An American Crusade for Wildlife* (New York: Winchester Press, 1975).

The relationship between nature and psychological health are the broad subject of the field of ecopsychology and its flagship journal, *Ecopsychology*. Laura Sewall and I explored "Why Ecopsychology Needs Natural History" in *Ecopsychology* 11 (2019): 78–80, and guest edited a special issue of the journal on "Reciprocal Healing: Nature, Health, and Wild Vitality," *Ecopsychology* 12, no. 3 (2020).

I explored the idea of "Natural History as a Practice of Kinship" in *Minding Nature* 12, no. 3 (2019): 12–15. The importance of kinship is considered more broadly in the five-volume *Kinship: Belonging in a World of Relations*, edited by Gavin Van Horn, Robin Wall Kimmerer, and John Hausdoerffer (Libertyville, IL: Center for Humans and Nature Press, 2021). My essay on natural history as kinship is in volume 5, *Practice*.

Terry Tempest Williams quoted Mother Teresa in *Finding Beauty in a Broken World* (New York: Vintage, 2008).

The idea that we are what we pay attention to is from: Thomas Lowe Fleischner, *Singing Stone: A Natural History of the Escalante Canyons* (Salt Lake City: University of Utah Press, 1999), xviii.

Barry Lopez discussed the importance of querencia in *The Rediscovery of North America*, (Lexington: University Press of Kentucky, 1990), pages unnumbered.

E. O. Wilson's quote on STEM education being the opposite of his own path is from his foreword to Piotr Naskrecki's *Hidden Kingdom: The Insect Life of Costa Rica* (Ithaca, NY: Cornell University Press, 2017), vi.

Templum is a Latin noun (the root of "temple") that has been defined as both an open space for observation, and as sanctuary or shrine. The blend of these two meanings hones in on my sense of the potential of any place we engage in the practice of natural history.

The quote by Annie Dillard is from *Pilgrim at Tinker Creek* (New York: Harper Perennial Modern Classics, 2013).

Wayne Shorter and Herbie Hancock's "Open Letter" is widely available online, including at: https://hollywoodawac.com/blogadmin/2018/12/21/wayne-shorter-amp-herbie-hancock -pen-an-open-letter-to-the-next-generation-of-artists.

"The naturalist's trance" was first described by Stephen Trimble in *Words From the Land: Encounters with Natural History Writing* (Layton, UT: Gibbs Smith, 1988); this term was the title of his introduction. The description by Robin Wall Kimmerer is from her essay, "Heal-All," in *Nature, Love, Medicine: Essays on Wildness and Wellness*, ed. T. L. Fleischner (Salt Lake City: Torrey House Press, 2017), 241–42.

Chapter 11: "Seeing, Feeling, and Knowing Nature," Crump

It is now known that other species of dendrobatids lack amplexus (the mating behavior in which the male clasps the female). The paper based on my observations in Costa Rica is Martha L. Crump, "Territoriality and mating behavior in *Dendrobates granuliferus* (Anura: Dendrobatidae)," *Herpetologica* 28 (1972): 195–98.

The quote from John Muir is from John Muir, *Our National Parks* (Cambridge, MA: Houghton, Mifflin, 1901). The quote from David Sobel is from David Sobel, *Beyond Ecophobia: Reclaiming the Heart of Nature Education* (Great Barrington, MA: Orion Society and the Myrin Institute, 1996). The quote from Liberty Hyde Bailey comes from Liberty Hyde Bailey, Leaflet I: "What Is Nature Study?," 1904.

For current data on fewer people spending time outdoors, see, for example, the *2020 Outdoor Participation Report*, Outdoor Foundation, December 31, 2020, https://outdoorindustry.org. For a discussion of the factors influencing conservationists and environmental educators to embrace their environmental values and philosophies, see Louise Chawla, "Significant Life Experiences Revisited: A Review of Research on Sources of Environmental Sensitivity," *Environmental Education Research* 4 (1998): 369–82. For the University of Vermont study of outdoor activity engagement during the COVID-19 pandemic see Joshua W. Morse, Tatiana M. Gladkikh, Diana M. Hackenburg, and Rachelle K. Gould, "COVID-19 and Human-Nature Relationships: Vermonters' Activities in Nature and Associated Nonmaterial Values during the Pandemic," *PLoS ONE* 15, no. 12 (2020): e0243697.

For human impact on land, see Erle C. Ellis et al., "People Have Shaped Most of Terrestrial Nature for at Least 12,000 Years," *PNAS* 118, no. 17 (2021): e2023483118. https://doi.org/10.1073 /pnas. For biodiversity on Indigenous lands, see Richard Schuster, Ryan R. Germain, Joseph R. Bennett, Nicholas J. Reo, and Peter Arcese, "Vertebrate Biodiversity on Indigenous-Managed Lands in Australia, Brazil, and Canada Equals That in Protected Areas," *Environmental Science and Policy* 101 (2019): 1–6. For discussions of protecting nature for nature's sake versus protecting nature for the benefit of people, see, for example, Tim Caro, Jack Darwin, Tavis Forrester, Cynthia Ledoux-Bloom, and Caitlin Wells, "Conservation in the Anthropocene," *Conservation Biology* 26

(2011): 185–88; Peter Kareiva and Michelle Marvier, "Conservation for the People," *Scientific American* 297 (2007): 50–57; Peter Kareiva, Robert Lalasz, and Michelle Marvier, "Conservation in the Anthropocene: Beyond Solitude and Fragility," *Breakthrough Journal* 2 (2011): 26–36; Peter Kareiva and Michelle Marvier, "What Is Conservation Science?," *BioScience* 62 (2012): 962–69; Emma Marris, *Rambunctious Garden: Saving Nature in a Post-Wild World* (New York: Bloomsbury, 2011); Emma Marris, Peter Kareiva, Joseph Mascaro, and Erle Ellis, "Hope in the Age of Man," *New York Times*, Op-Ed, December 7, 2011; various essays in Ben A. Minteer and Stephen J. Pyne, eds., *After Preservation* (Chicago: University of Chicago Press, 2015); Michael Soulé, "The 'New Conservation,'" *Conservation Biology* 27 (2013): 895–97.

Chapter 12: "Virtual Nature and the Future of the Wild," Clayton

Richard Louv's book, *Last Child in the Woods*, is published by Algonquin Books (updated edition, 2008). Peter Kahn has written extensively about the ways in which we respond to technology as compared to nature; see his 2011 book *Technological Nature: Adaptation and the Future of Human Life* from MIT Press. Bill McKibben's ideas were developed in *The End of Nature*, published in 2006 by Random House. The survey of *World of Warcraft* players (M. X. Truong, A. C. Prévot, and S. Clayton, "Gamers Like It Green: The Significance of Vegetation in Online Gaming," *Ecopsychology* 10, no. 1 [2018]: 1–13) can be found here: http://doi.org/10.1089/eco.2017.0037; the study showing that video gamers were learning about wild species was by Crowley, Silk, and Crowley (M. Crowley, E. Silk, and S. Crowley, "The Educational Value of Virtual Ecologies in *Red Dead Redemption 2*," *People and Nature* 3, no. 6 [2021]: 1229–43, https://doi.org/10.1002/pan3.10242). E. O. Wilson published his ideas about *Biophilia* in 1984 (Harvard University Press), though much more has been written since then. A review of the multisensory benefits of nature experiences (L. S. Franco, D. F. Shanahan, and R. A. Fuller, "A Review of the Benefits of Nature Experiences: More Than Meets the Eye," *International Journal of Environmental Research and Public Health* 14, no. 8 [2017]: 864) is available here: https://doi.org/10.3390/ijerph14080864.

Finally, some of these ideas have been discussed in S. Clayton, A. Colléony, P. Conversy, E. Maclouf, L. Martin, A. C. Torres, M. Truong, and A. C. Prévot, "Transformation of Experience: Toward a New Relationship with Nature," *Conservation Letters* 10, no. 5 (2017), 645–51, https://doi.org/10.1111/conl.12337; and in M.X.A. Truong and S. Clayton, "Technologically transformed experiences of nature: A challenge for environmental conservation?," *Biological Conservation* 244 (2020): 108532, https://doi.org/10.1016/j.biocon.2020.108532.

Chapter 13: "The Digital Animal," Adams

On Laikipia, its elephants, and conflicts with farmers, see: L. E. Evans and W. M. Adams, "Elephants as Actors in the Political Ecology of Human-Elephant Conflict," *Transactions of the Institute of British Geographers* 43 (2018): 630–45. For *Space for Giants*, see https://www.spaceforgiants.org/. On tracking elephants, see M. Graham, I. Douglas-Hamilton, P. C. Lee, and W. M. Adams, "The Movement of African Elephants in a Human-Dominated Land Use Mosaic," *Animal Conservation* 12 (2009): 445–55. For the animated tracks of crop-raiding elephants, see: "Space for Giants. Tracking Elephants Using Technology," https://www.youtube.com/watch?v=pOnOyjMo-uk.

For the history of radio tracking, see E. Benson, *Wired Wilderness: Technologies of Tracking and the Making of Modern Wildlife* (Baltimore, MD: Johns Hopkins University Press, 2010). The use of geolocators is described by E. Bächler, S. Hahn, M. Schaub, R. Arlettaz, L. Jenni, J. W. Fox, V. Afanasyev, and F. Liechti, "Year-Round Tracking of Small Trans-Saharan Migrants Using Light-Level Geolocators," *PloS ONE* 5 (2010): e9566. For examples of geolocators, see: "Products," Migrate Technology, Ltd., at https://www.migratetech.co.uk/geolocators_8.html. The Birds Canada Motus Wildlife Tracking System is described at: https://motus.org/. The research use of camera traps is described in J. McCallum, "Changing Use of Camera Traps in Mammalian Field Research: Habitats, Taxa and Study types," *Mammal Review* 43, no. 3 (2013): 196–206. The special issue of *Oryx* on camera trap studies was March 2021 (volume 55, issue 2). On satellites and penguin colonies in Antarctica, see P. T. Fretwell and P. N. Trathan, "Discovery of New Colonies by Sentinel-2 Reveals Good and Bad News for Emperor Penguins," *Remote Sensing in Ecology and Conservation* 7 (2020): 139–53. On drones for conservation, see L. P. Koh and S. A. Wich, *Conservation Drones: Mapping and Monitoring Biodiversity* (New York: Oxford University Press, 2018). On thermal imaging, see J. Witczuk, S. Pagacz, A. Zmarz, and M. Cypel, "Exploring the Feasibility of Unmanned Aerial Vehicles and Thermal Imaging for Ungulate Surveys in Forests—Preliminary Results," *International Journal of Remote Sensing* 39 (2018): 5504–21. On acoustic bat monitoring, see S. E. Newson, V. Ross-Smith, I. Evans, R. Harold, R. Miller, M. Horlock, and K. Barlow, "Bat-Monitoring: A Novel Approach," *British Wildlife* 25 (2014): 264–69; an example of an app to detect bats is Bat Recorder for Android (https://batmanagement.com/products/bat-recorder-app-for-android). On acoustic sensors to detect elephants, see C. M. Dissanayake, R. Kotagiri, M. N. Halgahmuge, and B. Moran, "Improving Accuracy of Elephant Localization Using Sound Probes," *Applied Acoustics* 129 (2018): 92–103. On marine acoustic tracking, see J. E. Stanistreet, D. Risch, and S. M. Van Parijs, "Passive Acoustic Tracking of Singing Humpback Whales (*Megaptera novaeangliae*) on a Northwest Atlantic feeding ground," *PLoS ONE* 8, no. 4 (2013), doi:10.1371/journal.pone.0061263.

On the different migratory paths taken by British common cuckoos, see C. M. Hewson, K. Thorup, J. W. Pearce-Higgins, and P. W. Atkinson, "Population Decline Is Linked to Migration Route in the Common Cuckoo," *Nature Communications* 7 (2016): 12296, doi:10.1038/ncomms12296. For a more general account of animal tracking, see J. Cheshire and O. Uberti, *Where Animals Go: Tracking Wildlife with Technology in 50 Maps and Graphics*, (London: Particular Books, 2016). For conservation applications of animal tracking, see T. Katzner and R. Arlettaz, "Evaluating Contributions of Recent Tracking-Based Animal Movement Ecology to Conservation Management," *Frontiers in Ecology and Evolution* 7 (2020), doi:10.3389/fevo.2019.00519; A. Verma, R. van der Wal, and A. Fischer, "Imagining Wildlife: New Technologies and Animal Censuses, Maps and Museums," *Geoforum* 75 (2016): 75–86; and E. S. Benson, "Trackable Life: Data, Sequence, and Organism in Movement Ecology," *Studies in History and Philosophy of Biological and Biomedical Sciences* 57 (2016): 137e147.

On hen harriers, see M. Murgatroyd, S. M. Redpath, S. G. Murphy, D.J.T. Douglas, R. Sauders, and A. Amar, "Patterns of Satellite Tagged Hen Harrier Disappearances Suggest Widespread Illegal Killing on British Grouse Moors," *Nature Communications* 10 (2019): 1094, https://doi.org/10.1038/s41467-019-09044-w. For sentinel albatrosses, see: H. Weimerskirch, J. Collet, A. Corbeau, A. Pajot, F. Hoarau, C. Marteau, D. Filippi, and S. C. Patrick, "Ocean Sentinel Albatrosses Locate

Illegal Vessels and Provide the First Estimate of the Extent of Nondeclared Fishing," *Proceedings of the National Academy of Sciences* 117 (2020): 3006–14. The tagged great white shark killed in Western Australia is discussed by J. J. Meeuwig, R. Harcourt, and F. G. Whoriskey in "When Science Places Threatened Species at Risk," *Conservation Letters* 8 (2015): 151–52.

On charismatic species, see J. Lorimer, "Nonhuman Charisma," *Environment and Planning D: Society and Space* 25 (2007): 911–32; and A. Verma, R. van der Wal, and A. Fischer, "Microscope and Spectacle: On the Complexities of Using New Visual Technologies to Communicate about Wildlife Conservation," *Ambio* 44, Suppl. 4 (2014): 648, doi:10.1007/s13280-015-0715-z. On the record-breaking migration: D. Boffey, "'Jet Fighter' Godwit Breaks World Record for Non-Stop Bird Flight," *The Guardian*, October 13, 2020, https://www.theguardian.com /environment/2020/oct/13/jet-fighter-godwit-breaks-world-record-for-non-stop-bird-flight. On the RSPB Osprey tracking project, see: "Tracking Ospreys," RSPB, https://www.rspb.org .uk/our-work/conservation/satellite-tracking-birds/tracking-ospreys/. The BTO cuckoo tagging program is described here: "Cuckoo Tracking Program," British Trust for Ornithology, https://www.bto.org/our-science/projects/cuckoo-tracking-project, and the sponsorship opportunity here: "Sponsor a Cuckoo," British Trust for Ornithology, https://www.bto.org/our -science/projects/cuckoo-tracking-project/get-involved/sponsor-cuckoo.

On technology and the woods, see Richard Louv, *Last Child in the Woods: Saving Our Children from Nature-Deficit Disorder* (Chapel Hill, NC: Algonquin Books, 2005). The Colin Ellard quote is in *Places of the Heart: The Psychogeography of Everyday Life* (New York: Bellevue Literary Press, 2015), 193.

Chapter 14: "Hope for the Wild in Afrofuturism," Schell

On the biodiversity impact of the Anthropocene, see Georgina Gustin and John H. Cushman Jr., "Humanity Faces a Biodiversity Crisis. Climate Change Makes It Worse," *Inside Climate News*, November 30, 2020, https://insideclimatenews.org/news/06052019/climate-change -biodiversity-united-nations-species-extinction-agriculture-food-forests/. For the most recent IPCC assessments of climate change, see IPCC, *Climate Change 2022: Mitigation of Climate Change. Contribution of Working Group III to the Sixth Assessment Report of the Intergovernmental Panel on Climate Change*, ed. P. R. Shukla, J. Skea, R. Slade, A. Khourdajie, R. van Diemen, D. McCollum, M. Pathak, S. Some, P. Vyas, R. Fradera, M. Belkacemi, A. Hasija, G. Lisboa, S. Luz, J. Malley (Cambridge, UK, and New York: Cambridge University Press, 2022), doi:10.1017/9781009157926; and Lauren Sommer, "It's Not Too Late to Stave Off the Climate Crisis, U.N. Report Finds. Here's How," National Public Radio (NPR), April 4, 2022, https:// www.npr.org/2022/04/04/1090577162/climate-change-un-ipcc-report.

Ecological and climate grief are discussed in John S. Dryzek, Richard B. Norgaard, and David Schlosberg. *Climate Change and Society: Approaches and Responses* (Oxford University Press, 2011), https://doi.org/10.1093/oxfordhb/9780199566600.003.0001; and Ashlee Cunsolo and Neville R. Ellis, "Ecological Grief as a Mental Health Response to Climate Change–Related Loss," *Nature Climate Change* 8, no. 4 (2018): 275–81, https://doi.org/10.1038/s41558-018-0092-2.

The Black Joy Parade is featured at https://www.blackjoyparade.org/. For more on Odums, see Katy Reckdahl and Akasha Rabut, "This New Orleans Artist Challenges the Way People

See Things," *New York Times*, March 9, 2020, https://www.nytimes.com/2020/03/09/arts /bmike-artist-new-orleans.html. Michelle Obama's story is recounted in Brandee Sanders, "'Journey to White House Took Multiple Lifetimes': Ava DuVernay Pays Homage to Michelle Obama," NewsOne, August 6, 2017. https://newsone.com/3730366/ava-duvernay-michelle -obama-slave-ancestor/.

The significance of Afrofuturism, its definitions, and its centrality to dreaming of a better reality is articulated by Claire Elise Thompson in "From Afrofuturism to Ecotopia: A Climate-Fiction Glossary," Fix Solutions Lab, September 14, 2021, https://grist.org/fix/climate-fiction /afrofuturism-to-ecotopia-climate-fiction-glossary/.

The Afrofuturist dimensions of the Black Panther comic are explored in Myron T. Strong and K. Sean Chaplin, "Afrofuturism and Black Panther," *Contexts* 18, no. 2 (2019): 58–59, https:// doi.org/10.1177/1536504219854725. John Lewis's remarks ("Freedom is not a state") appear in his book *Across That Bridge: A Vision for Change and the Future of America* (New York: Hachette Books, 2016).

On the pseudonym "Oakanda," see Josie Clerfond, "Oakanda—The Guide to California's East Bay," Spirited Pursuit, September 14, 2018, https://www.spiritedpursuit.com/blog /oakandathe-guide-to-californias-east-bay.

On the history of Black Americans, see Nikole Hannah-Jones, *The 1619 Project: A New Origin Story* (New York: One World, 2021).

On the influence of past racial segregation on urban ecology, see Steward T. A. Pickett and J. Morgan Grove, "An Ecology of Segregation," *Frontiers in Ecology and the Environment* 18, no. 10 (2020): 535, https://doi.org/10.1002/fee.2279. For work examining the impact of racial inequality on genetic diversity and species richness, see Chloe Schmidt and Colin J. Garroway, "Systemic Racism Alters Wildlife Genetic Diversity," *Proceedings of the National Academy of Sciences* (2022), https://doi.org/10.1073/pnas.2102860119; and Dexter H. Locke, Billy Hall, J. Morgan Grove, Steward T. A. Pickett, Laura A. Ogden, Carissa Aoki, Christopher G. Boone, and Jarlath P. M. O'Neil-Dunne, "Residential Housing Segregation and Urban Tree Canopy in 37 US Cities," *Npj Urban Sustainability* 1, no. 1 (2021): 15, https://doi.org/10.1038/s42949-021-00022-0. The role of redlining on urban ecological systems is discussed in Morgan Grove, Laura Ogden, Steward Pickett, Chris Boone, Geoff Buckley, Dexter H. Locke, Charlie Lord, and Billy Hall, "The Legacy Effect: Understanding How Segregation and Environmental Injustice Unfold over Time in Baltimore," *Annals of the American Association of Geographers* 108, no. 2 (2018): 524–37, https://doi.org/10.1080 /24694452.2017.1365585; and Christopher J. Schell, Karen Dyson, Tracy L. Fuentes, Simone Des Roches, Nyeema C. Harris, Danica Sterud Miller, Cleo A. Woelfle-Erskine, and Max R. Lambert, "The Ecological and Evolutionary Consequences of Systemic Racism in Urban Environments," *Science* 369, no. 6510 (2020): eaay4497, https://doi.org/10.1126/science.aay4497. For work examining biodiversity sampling and redlining, see Diego Ellis-Soto, Melissa Chapman, and Dexter H. Locke, (2022). "Uneven Biodiversity Sampling across Redlined Urban Areas in the United States," *EcoEvoRxiv Preprints*, https://ecoevorxiv.org/repository/view/3736/.

The principle of the First National People of Color Environmental Leadership summit are described in Paul Mohai, David Pellow, and J. Timmons Roberts, "Environmental Justice," *Annual Review of Environment and Resources* 34, no. 1 (2009): 405–30, https://doi.org/10.1146/annurev -environ-082508-094348. The One Health framework is discussed in Ronald M. Atlas, "One

Health: Its Origins and Future," in *One Health: The Human-Animal-Environment Interfaces in Emerging Infectious Diseases*, ed. John S. Mackenzie, Martyn Jeggo, Peter Daszak, and J. Richt (Berlin: Springer Berlin Heidelberg, 2012), 1–13, https://doi.org/10.1007/82_2012_223.

On racial capitalism, see Cedric J. Robinson, *Black Marxism: The Making of the Black Radical Tradition* (Chapel Hill: University of North Carolina Press, 2000).

On the use of Adobe mud blocks in inno-native design in Accra from designer Joe Osae-Addo, see Frances Anderton, "An Inno-native Approach," Dwell, January 16, 2009, https://www.dwell.com/article/an-inno-native-approach-adefecc6.

For context on climate "doomism" and guarding against it, see Marco Silva, "Why Is Climate 'Doomism' Going Viral—And Who's Fighting It?," BBC News, May 23, 2022. https://www.bbc.com/news/blogs-trending-61495035.

For coverage on the Bay Area, California's summer fish die-off, see Terry Chea and Olga R. Rodriguez, "Thousands of Dead Fish in San Francisco Bay Area Blamed on Toxic Red Tide," *Los Angeles Times*, September 1, 2022, https://www.latimes.com/california/story/2022-09-01/dead-fish-in-san-francisco-bay-area-blamed-on-toxic-red-tide. For coverage on the historic deluge of atmospheric rivers in California, see Aaron Scott, Thomas Lu, and Gabriel Spitzer, "An Atmospheric River Runs Through It," NPR-KQED, January 6, 2023. https://www.npr.org/2023/01/04/1146963688/an-atmospheric-river-runs-through-it.

Editorial on the symbolism of *Wakanda Forever* Shuri dialogue by Marco Vito Oddo, "This One Scene in 'Civil War' Sets Up the Entire Premise of 'Wakanda Forever,'" Collider, November 13, 2022, https://collider.com/civil-war-black-panther-wakanda-forever-revenge-is-consuming-you/.

Chapter 15: "Listening to Learn: Nature's Hot and Cold Extremes," Berger

Specific references to many of the accounts from Namibia can be found in C. Cunningham and J. Berger, *Horn of Darkness, Rhinos on the Edge* (New York, Oxford University Press, 1997). For the Arctic and muskoxen accounts with Freddy Goodhope references are in J. Berger, *Extreme Conservation, Life at the Edges of the World* (Chicago: University of Chicago Press, 2018). The quote from Bab Dioum can be found at: https://www.goodreads.com/quotes/6430296-in-the-end-we-will-conserve-only-what-we-love. Information on Fat Bear Week is available at: https://www.nps.gov/katm/learn/fat-bear-week.htm; and for projected changes in land use in South America, see https://population.un.org/wpp/.

Afterword: "A Part or Apart: Ought Nature Lovers Ever Wear Fur?," Greene

Epigraph quoted from Rita Hosking's "Resurrection," on the album *Frankie and the No-Go Road* (2015), and used with her permission.

My PhD advisor and a fellow grad student on animals and values: G. M. Burghardt, and H. A. Herzog Jr., "Beyond Conspecifics: Is Brer Rabbit Our Brother?" *BioScience* 30 (1980): 763–68.

"The Answer," in *The Selected Poetry of Robinson Jeffers*, ed. T. Hunt (Palo Alto, CA: Stanford University Press, 2002).

Background: H. W. Greene, *Tracks and Shadows: Field Biology as Art* (Berkeley: University of California Press, 2013); H. W. Greene, "Pomegranates, Peccaries, and Love," *Ecopsychology* 12 (2020): 166–72; D. M. Hillis, *Armadillos to Ziziphus: A Naturalist in the Texas Hill Country* (Austin: University of Texas Press, 2023); L. M. Weber, *Understanding Nature: Ecology for a New Generation* (Boca Raton, FL: CRC Press, 2023).

Fur: J. Oakes et al., "Comparisons of Manufactured and Traditional Cold Weather Ensembles," *Climate Research* 5 (1995): 83–90; A. J. Cotel et al., "Effects of Ancient Inuit Fur Parka Ruffs on Facial Heat Transfer," *Climate Research* 26 (2004): 77–84; I. Verheijgen et al., "Early Evidence for Bear Exploitation during MIS 9 from the Site of Schönigen 12 (Germany)," *Journal of Human Evolution* 77 (2023): e103294.

Humans in nature: T. C. Rick et al., "Ecological Change on California's Channel Islands from the Pleistocene to the Anthropocene," *BioScience* 64 (2014): 680–92; E. C. Ellis et al., "People Have Shaped Most of Terrestrial Nature for at Least 12,000 Years," *Proceedings of the National Academy of Sciences* 118 (2021): e2023483118; L. R. Burger, "Predatory Bird Damage to the Taung Type Skull of *Australopithecus africanus* Dart 1925," *American Journal of Physical Anthropology* 131 (2006): 166–68.

Last Texas jaguar: Scott Dubois, "The Last Jaguar in Texas," Wild Texas History, April 11, 2020, accessed June 3, 2023, https://wildtexashistory.com/the-last-jaguar-in-texas-1948/.

Microplastics: X. Z. Lim, "Microplastics Are Everywhere—But Are They Harmful?," *Nature* 593 (2021): 22–25.

Thoughtful case against humans using animals: B. J. King, *Personalities on the Plate: The Lives and Minds of Animals We Eat* (Chicago: University of Chicago Press, 2017).

Indigenous values: E. Salmón, *Eating the Landscape: American Indian Stories of Food, Identity, and Resilience* (Tucson: University of Arizona Press, 2012); R. W. Kimmerer, *Braiding Sweetgrass: Indigenous Wisdom, Scientific Knowledge, and the Teachings of Plants* (Minneapolis: Milkweed Editions, 2015); C. A. Hofman et al., "Collagen Fingerprinting and the Earliest Marine Mammal Hunting in North America," *Science Advances* 8 (2018): 10014; M. L. Moss, "Did Tlingit Ancestors Eat Sea Otters? Addressing Intellectual Property and Cultural Heritage through Zooarchaeology," *American Antiquity* 85 (2020): 202–21.

Raccoons: L. A. Nell et al., "Presence of Breeding Birds Improves Body Condition for a Crocodilian Nest Protector," *PLoS ONE* 11, no. 3 (2016): e0149572; O. Hernández-Gallegos et al., "Massive Predation of Pine Toad, *Incilius occidentalis* (Anura: Bufonidae)," *Caldasia* 41 (2019): 450–52.

Human exceptionalism: R. Sussman and L. Sussman, "Of Human Nature and on Human Culture: the Importance of the Concept of Culture in Science and Human Society," in *Verbs, Bones, and Brains: Interdisciplinary Perspectives on Human Nature*, ed. A. Fuentes and A. Visala (Notre Dame, IN: University of Notre Dame Press, 2017), 58–69; B. Swartz and B. D. Mishler, *Speciesism in Biology and Culture: How Human Exceptionalism is Pushing Planetary Boundaries* (Cham, Switzerland: Springer, 2022).

Philosophical differences: C. Diehm, "Rolston on Animals, Ethics, and the Factory Farm," *Expositions* 6, no. 1 (2012): 29–40; J. McMahan, (2010), "The Meat Eaters," *New York Times Opinionator*, accessed January 18, 2023, https://archive.nytimes.com/opinionator.blogs.nytimes.com/2010/09/19/the-meat-eaters/; B. A. Minteer, *Refounding Environmental Ethics: Pragmatism, Principle, and Practice* (Philadelphia: Temple University Press, 2012).

Conservation: D. H. Janzen, "Gardenification of Wildland Nature and the Human Footprint," *Science* 279 (1998): 344–45; F. Pierce, "Sparing versus Sharing: The Great Debate over How to Protect Nature," Yale 360, 2018, accessed January 18, 2023, https://e360.yale.edu/features/sparing-vs-sharing-the-great-debate-over-how-to-protect-nature; F. Keesing et al., "Consequences of Integrating Livestock and Wildlife in an African Savanna," *Nature Sustainability* 1 (2018): 561–73; A. D. Manning et al., "Stretch Goals and Backcasting: Approaches for Overcoming Barriers to Large-Scale Ecological Restoration," *Restoration Ecology* 14 (2006): 487–92.

Jaguars and ranching: I. Cassainge et al., "Augmentation of Natural Prey Reduces Cattle Predation by Puma (*Puma concolor*) and Jaguar (*Panthera onca*) on a Ranch in Sonora, Mexico," *Southwestern Naturalist* 65 (2021): 123–30; Mexican ranchers reported horned cattle are less preyed upon, I. Cassaigne, email to author.

INDEX

Page numbers in *italics* indicate figures and tables.